ADVANCES IN MOTOR TORQUE CONTROL

Edited by **Mukhtar Ahmad**

Advances in Motor Torque Control
Edited by Mukhtar Ahmad

Published by InTech
Janeza Trdine 9, 51000 Rijeka, Croatia

Technical Editor Sonja Mujacic

Cover Designer InTech Design Team

Additional hard copies can be obtained from orders@intechopen.com

Advances in Motor Torque Control,
Edited by Mukhtar Ahmad
p. cm.
ISBN 978-953-307-686-7

For EU product safety concerns:
IN TECH d.o.o., Prolaz Marije Krucifikse Kozulić 3, 51000 Rijeka, Croatia,
info@intechopen.com or visit our website at intechopen.com.

Meet the editor

Dr. Mukhtar Ahmad is currently working as professor of electrical engineering at Z H College of Engineering, Aligarh Muslim University, Aligarh. He is a senior member of IEEE. He obtained his B.Sc in engineering in 1969, M.Sc in engineering in 1972, and Ph.D in 1991, all from Aligarh Muslim University, Aligarh. He worked as associate professor at University Putra Malaysia for two years and at Multimedia University Cyberjaya for one year. He is author of two books on electric drives. His book "High Performance AC drives" was published in 2010. He has published more than 30 papers in international journals in the area of power systems, power electronics and electrical drives.

Contents

Preface

More than fifty percent of total electrical energy generated is converted into mechanical energy with the help of electric motors. If the efficiency of these motors is improved, a large amount of energy can be saved. With the advent of power electronics it is now possible to control the torque and speed of electric motors precisely as needed, resulting in saving of energy.

In this book various methods of control of electric motors are discussed. Sensorless control of motors requires the estimation of speed without actually measuring it. Similarly now the torque is controlled directly by the direct torque control method described in the book. Reluctance motors are also finding applications in many industries like consumer appliances, automobiles and defense. Special application of motors in dentistry is also described. The effect of human reaction on motor performance is also described.

Dr. Mukhtar Ahmad,
senior member IEEE Professor,
Department of Electrical Engineering,
Aligarh Muslim University,
India

1

Intelligent Torque Control Motors in Dentistry

Mohammad Hassan Zarrabi and Maryam Javidi
Endodontic Department, Dental School,
University of Medical Sciences, Mashhad,
Iran

1. Introduction

Perhaps no instrument has more optimized in dentistry and the dentist over the past century in the mind of the public than the dental headpiece. Prior to 1870 dentists had no driven rotary tools. During the 1850-1870 periods various other instruments were advised to rotate burs in cavities.

Early example of clock wise drill was patented in 1864, come into use by 1871, It was with the advent of the foot engine that the first dental hand pieces came into being.

Straight hand pieces with a variety of intricate chuck-closing mechanisms became well developed during the 1880, and since they were permanently linked to the foot - engine flexible cable were converted into angle hand pieces by connecting so - called " lock - bit attachments" to their front ends, these lock-bits being available in right angle, acute angle and obtuse angle patterns.(Fig 1)

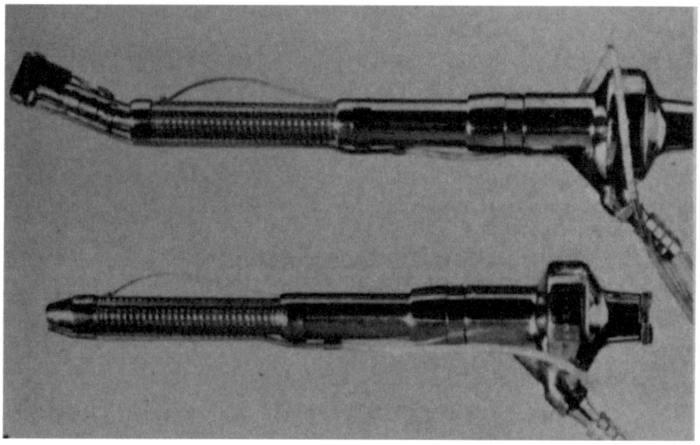

Fig. 1. Dental hand piece in 19th century.

From 1875 onwards, the use of the foot engine became widespread, but its demise was foreshadowed by the advent of electric headpiece driving mechanism.

Early electric motors were designed to be attached to foot engines or alternatively as independent entities.

In briefly, dental hand pieces are small, very specialized air as electric driven turbines used in both high and low speed dental hand pieces.

2. Structure of dental hand piece; turbine

A turbine is typically made up of multiple components including:
Rotor: Integral shaft on which components are mounted
Chuck: usually housed within the rotor.
Impeller: Component which converts energy from pressurized air or electricity in to the rotational motion of the turbine required for cutting.
Bearing: Allow entire assembly to spin freely with as little friction as possible.
Rings: Provide firm seating inside hand piece head while minimizing vibration.
High speed dental hand pieces turn at approximately 400/000 revolutions per minute (rpm).
Slow speed dental hand piece turn at 150 rpm to 2000 rpm.(Fig 2)

Fig. 2. Structure of dental hand piece.

Because of the specialized nature, very high speed, the turbines that drive them must be manufactured to the highest standards. Standard materials would result in imminent failure, and possible harm to the patient on which the equipment was being used.
Hand piece manufacturers sometimes design hand pieces around the mast efficient, highest power producing dental hand piece turbine that current technology allows.
Speed and torque is king where dental hand pieces are concerned and competition is fierce between manufacturers.

3. What is torque?

Torque is a measure of how much force acting on an object cause that object to rotate.
Torque also called moment ox moment of force is a tendency of force to rotate an object on axis, fulcrum, or pivot, just as a force is a push or a pull, a torque can be thought of as a twist.
In simple terms, torque is a measure of the turning force on an object such as a bolt oz a fly wheel, for example, pushing oz pulling the handle of a wrench connected to a nut or bolt produces a torque (turning force) that loosens or tightens the nut or bolt. (Fig 3)

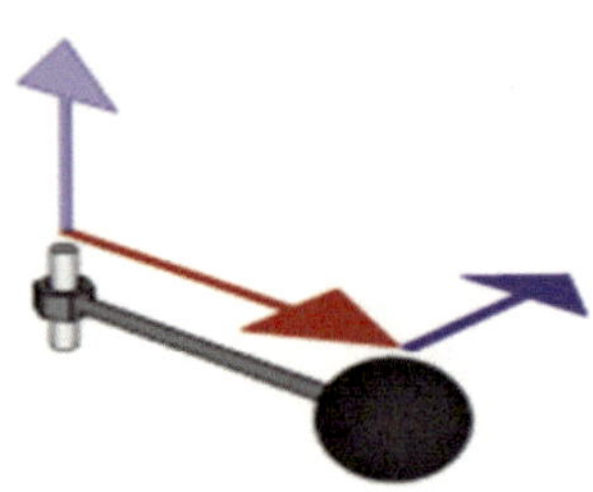

Fig. 3. Turning force.

4. Torque in dentistry; root canal therapy

In many aspect of practice in dentistry, especially in root canal therapy for root canal preparation, there is a turning force on an instrument.
Torque is a parameter that must be controllable in root canal preparation, because of different instruments which have been used, seem to need different values of torque.
In root canal preparation, safety usage of instrument depends on considering the torque at failure of instrument.
The instruments are subjected to different of torsional torque, if the level of torque is equal to or greater than the torque at failure (fracture), the instrument will separate.

5. Torque control hand pieces

Different types of hand pieces are used in conjunction with the rotary instruments, the air and electric motors without torque control and the electric torque control motors. (Fig 4)

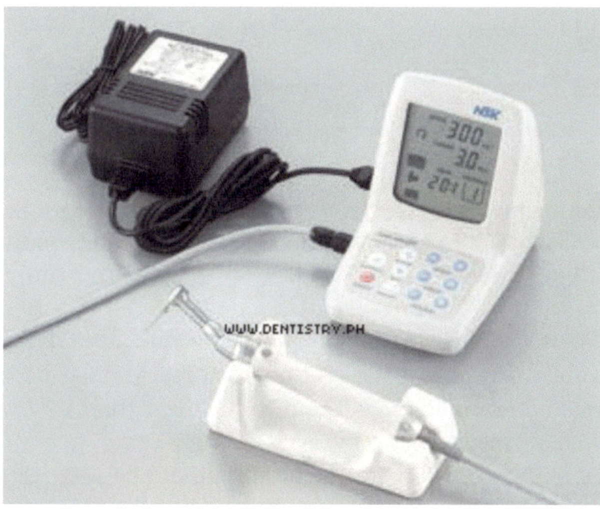

Fig. 4. Torque control motor.

Theoretically, the torque control hand pieces (motors) take into consideration the torque at failure of rotary instrument.
Torque values lower than the torque at fracture of the instruments can be set on the torque control hand pieces.
When a high torque control hand pieces is used the instrument is very active and the incidence of instrument locking and, consequently, deformations and separation would tend to increase.
Air driven hand pieces or air motors do not allow torque control and variation in air pressure could affect the rotational speed and, consequently, torque. For instance a drop in air pressure would lead to a decrease of torque. (Fig 5)
The instrument would become less active, and the operator would tend to force the instrument in to the canal of teeth leading to deformation and separation.

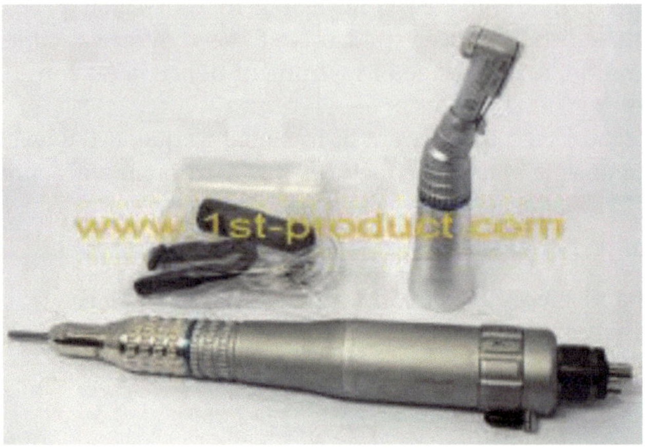

Fig. 5. Air driven hand piece.

Recently a generation of low and very low torque control motors has been introduced; torque values as low as $1\ {}^{N}/_{cm^2}$ can be set on these torque control motors, respectively, these motors take into consideration and low torque at failure values of rotary instruments.
If the high-torque is used the instrument specific torque limit is often exceeded, thus increasing the mechanical stress and the risk of fractures, it must be emphasized that the elastic limit of the tested instrument was found to be lower than $1\ {}^{N}/_{cm^2}$ when subjected to torsional testing.
To limit this potential breakage, a low torque motor should be used , if the torque is set just below the limit of elasticity for each instrument, the mechanical stress is lower, the risk of deformation and separation is likely to be reduced to an extent far below what has been possible before.
With the low torque motor, the motor will stop from rotating and can even reverse the direction of rotate when the instrument is subjected to torque level equal to the torque value set on the motor thus instrument failure would be avoided.

6. Different types of torque control electro motors

NSK Brasseler ENDO-MATE DT, ENDO-MATE DT is specifically designed for use with Ni-Ti files from all major suppliers. User-programmable preset memory can store up to 9 speed and torque settings exactly to the supplier's spec. A compact and lightweight control unit offers convenience of full portability between offices. ENDO-MATE DT can be hooked up directly to wall outlet or used with rechargeable battery. A large LCD display offers higher visibility for instantaneous recognition of micromotor status (Fig 6).

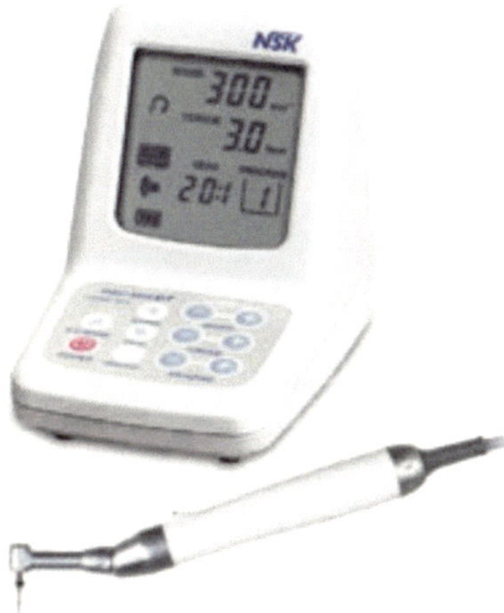

Fig. 6. ENDO-MATE DT.

J Morita root ZX® II low speed handpiece module

Root ZX II can easily be upgraded to a low speed handpiece offering speeds from 150 - 800 rpms. The low speed handpiece module is interchangeable and snaps easily onto the back of the unit. This new versatility allows the clinician to choose between apex locator, low speed handpiece, or a combination of both. Designed for enhanced performance, the tailor-made handpiece is lightweight (70 g) and has a compact head height (12.5 mm). Proven Root ZX II technology delivers extreme accuracy and reliability, while the display screen allows the clinician to visualize file movement during instrumentation (Fig 7).

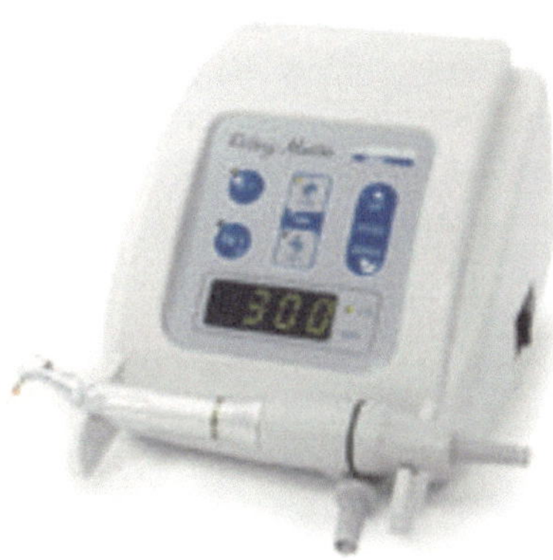

Fig. 7. J Morita root ZX® II.

J Morita Tri Auto ZX. The Root ZX II low speed handpiece is loaded with automatic safety functions. A new feature, Auto Torque Slow Down, offers added protection when preparing the canal. The file automatically slows down as the torque load approaches its set limit helping to reduce file breakage (Fig. 8).

Fig. 8. J Morita Tri Auto ZX.

The cordless Tri Auto ZX is the only endodontic handpiece with a built-in apex locator, providing the capability and convenience to electronically monitor the root canal before, during and after instrumentation. With the combined technology and accuracy of the Root ZX apex locator, the Tri Auto ZX can significantly increase accuracy and safety. The Tri Auto ZX also offers greater control and flexibility with the adjustable torque settings. The choice of automatic or manual mode operations provide versatility. The three automatic functions include: Auto Start/Stop, Auto Apical Reverse and Auto Torque Reverse.

J Morita rotary master® electric low speed motor

The Rotary Master is a lightweight, ergonomically designed, low speed electric motor with a consistent operating speed, regardless of the load applied to the rotary file. It is a perfect complement to any nickel titanium rotary file system. The Rotary Master comes with a 16:1 contra angle and boasts one of the smallest contra heads on the market. The variable speeds allow the unit to be used for a wide variety of general and endodontic procedures. Other features include a large digital rpm display, touch-panel adjustments and a motor reverse. The 1:1 contra angle and automatic crown and bridge remover are optional (Fig. 9).

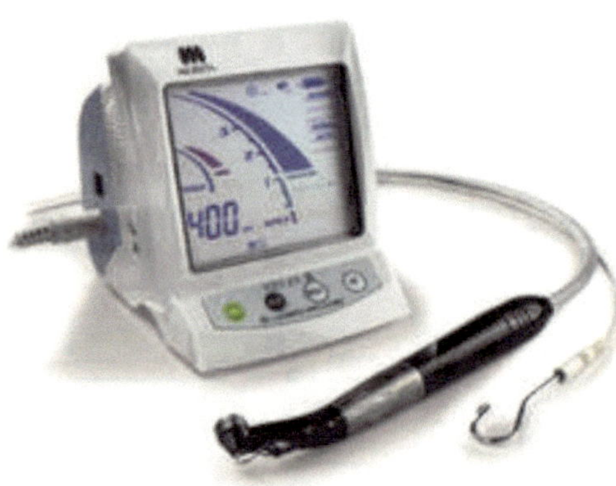

Fig. 9. J Morita rotary master® electric low speed motor.

TCM Endo 3The microprocessor controlled TCM Endo III is a slow-speed, electric torque-control motor capable of achieving faster and easier root canal preparation. Speed and maximum torque levels are preselectable and constantly controlled by the TCM III control unit. Speed is constant until the adjusted torque limit is reached, then the motor will reverse for 2 revolutions and return to the forward direction to finish root canal preparation. The TCM III is compatible with both Quantec and K3 Rotary Systems (Fig. 10).

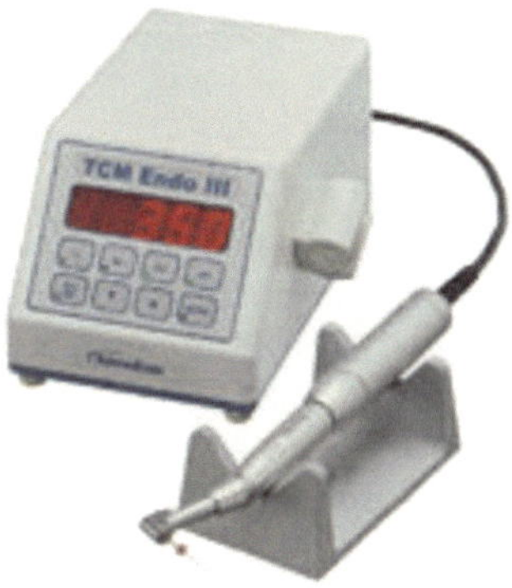

Fig. 10. TCM Endo III.

7. References

Richard F, Stephen. The dental handpiece-a history of its development. Aust Dent J. 1986; 31:165-180.

D. Gekelman, R. Ramamurtby, S. Mirfarsi, F. Paque, A. Peters. Rotary nickel-titanium GT and ProTaper files for root canal shaping by novice operators: A radiographic and micro-computed tomography evaluation. J Endod 2009; 35: 1584-1588.

A. Guelzow, O. Stamm, P. Martus, AM. Kielbasas. Comparative study of six rotary nickel–titanium systems and hand instrumentation for root canal preparation. Int Endod J. 2005; 38(10): 743-752.

M. Kuzekanani, L J Walsh, M A Yousefi. Cleaning and shaping curved root canals: Mtwo vs ProTaper instruments, a lab comparison. Indian J Dent Res 2009; 20: 268-70.

B. Jodway, M. Hulsmann. A comparative study of root canal preparation with NiTi-TEE and K3 rotary Ni-Ti instruments. Int Endod J. 2006; 39: 71-80.

Peters OA, Koka RS. Preparation of coronal and radicular spaces. In: Ingle JI, Bakland LK, Baumgartner JC. Endodontics. 6th ed. Hamilton: BC Decker Inc. 2008: 877-991.

Schneider SW. A comparison of Canal preparation in straight and curved root canals. Oral Surg Oral Med Oral Pathol 1971; 32: 271-5.

Short JA, Morgan LA, Baumgartner JC. A comparison of the effects on canal transportation by four instrumentation techniques. J Endod. 1997; 23: 503-7.

Garip Y. Gunday M. The use of computed tomography when comparing NiTi and SS File duving preponation of simulated curved canals. Int Endod J 2001; 34: 45-57.

Schafer E, Erler M, Dammaschke T. Comparative study of the shaping ability and cleaning efficiency of rotary Mtwo instruments: Part1: Shaping ability in simulated curved canals. Int Endod J 2006; 39: 196-202.

Schafer E, Erler M, Dammaschke T. Comparative study of the shaping ability and cleaning efficiency of rotary Mtwo instruments: Part2: Cleaning effectiveness and shaping ability in severely curved root canals of extracted teeth. Int Endod J 2006; 39: 203-12.

Gambarini G. Advantages and disadvantages of new torque-controlled endodontic motors and low-torque NiTi rotary instrumentation. Aust Endod J. 2001 Dec; 27(3): 99-104.

Hülsmann M, Stryga F. Comparison of root canal preparation using different automated devices and hand instrumentation. J Endod. 1993 Mar; 19(3): 141-5.

Patiño PV, Biedma BM, Liébana CR, Cantatore G, Bahillo JG. The influence of a manual glide path on the separation rate of NiTi rotary instruments. J Endod. 2005 Feb; 31(2): 114-6.

Li UM, Lee BS, Shih CT, Lan WH, Lin CP. Cyclic fatigue of endodontic nickel titanium rotary instruments: static and dynamic tests. J Endod. 2002 Jun; 28(6): 448-51.

Schäfer E, Diez C, Hoppe W, Tepel J. Roentgenographic investigation of frequency and degree of canal curvatures in human permanent teeth. J Endod. 2002 Mar; 28(3): 211-6.

Bramante CM, Bebert A, Barges RP. A methodology for evaluation of root canal instrumentation. J Endodon 1987; 13: 243-245.

Sonntag D, Kook K. Root canal preparation with NiTi systems K3, Mtwo and Protaper. Aust Endod J 2007; 33: 73-81.

Merret SJ, Bryant ST, Dummer PM.Comparison of the shaping ability of RaCe and FlexMaster rotary nickel-titanium system in simulated canals. J Endod 2006 Oct; 32(10): 960-2.

Javaheri H, Javaheri GH. A comparison of three NiTi rotary instruments in apical transportation. J Endod 2007; 33: 284-6.

Part 2

Human Factor and Reactive Torque Control

2

Experimental Verification of Reaction Torque Control Based on Driver Sensitivity to Active Front Steering

Ryo Minaki and Yoichi Hori
The University of Tokyo
Japan

1. Introduction

In today's world, the effectiveness of electronic active control systems in stabilizing a vehicle's motion has been recognized; thus, numerous active control systems have been developed to realize effective braking torque and traction control and have been incorporated in mass-produced vehicles. However, an effective active control system for steering is not yet available. Active front steering (AFS) is an effective technique to stabilize the motion control when methods such as the Direct Yaw Control (DYC) are used; however, for a vehicle, it is difficult to resolve the problem of interference between driver steering and automatic steering with AFS. Some studies have verified that AFS is effective from the viewpoint of vehicle motion physics but such verification alone is insufficient. Since the driver has control of the steering wheel, driver sensitivity interferes with vehicle motion control, and the theoretical effect of control by AFS cannot be realized. This problem is known as the steering interference problem. Therefore, it is important to ensure that the active steering control system does not interfere with driver steering. In this paper, we evaluate driver sensitivity quantitatively using a steering device, and we verify the efficacy of the reaction torque control method based on the fundamental characteristics of driver sensitivity.

2. Active front steering

In traditional steering systems, due to the presence of mechanical parts such as the torsion bar spring and the intermediate shaft between the steering wheel and front axle wheel, the inclinations of both wheels are directly related to each other. For this reason, it is not possible to realize AFS in such conventional systems. Either of two techniques can be used to solve this mechanical problem: steer-by-wire (SBW) and differential steering. SBW allows the steering wheel and front wheels to be controlled independently by replacing mechanical units with electric signal lines. On the other hand, differential steering controls the differential angle between the steering wheel and front axle wheels using a particular gear such as a planetary gear or harmonic gear. However, with these techniques, the following problems must be overcome to realize AFS. Experimental verification particularly describes the problems at the chapter 4.

2.1 Problems with AFS when using a differential gear

Since this technique links the steering wheel to the front axle via a differential gear, the reaction torque of the disturbance from the road surface is directly transmitted to the driver. For this reason, this technique inevitably causes steering interference when AFS intervenes strongly during steering.

2.2 Problems with AFS when using Steer-By-Wire

Since this technique does not link the steering wheel to the front axle via mechanical components, the reaction torque from the road surface is not transmitted to the driver at all. For this reason, the technique can decouple the steering interference completely. This is an advantage for AFS. However, when the front wheels hit a bump in the road, or if the driver operates the vehicle on gravel, this technique cannot transmit the reaction torque as road information to the driver. This is a critical problem for safe driving.

For realizing an effective AFS system, it is important to control the reaction torque transmitted to the driver for safe operation. The purpose of our study was to evaluate driver sensitivity quantitatively, to propose novel techniques for controlling the reaction torque based on driver sensitivity, and to verify whether these techniques were effective by using a steering device.

2.3 Active front steering control

A planetary gear structure is shown in Fig. 1. It consists of sun gear, ring gear and carrier. Each gear is conected to steering axle, front wheel axle, and sub motor. The block diagram of a steering system using a planetary gear for AFS is shown in Fig. 2. The block diagram of the AFS control is shown in Fig. 3. State variables for this system are shown in Table 1. The angle equation is shown in (1). α is gear ratio determined by number of gear teeth. The sub motor controls the ring gear angle in the planetary gear shown in (2) for AFS. The AFS sets the angle $\Delta\theta$ between the steering wheel and the front axle wheel based on vehicle motion controller. As a result, the AFS controls the front axle wheel angle shown in (3). The equation is calculated using the equations (1) and (2). The main motor is used to reduce the steering load in order to assist the driver. The toque equation of the planetary gear is shown in (4). The torque is not relationship between the gear angle.

$$\theta_m = -\frac{\theta_f}{\alpha} + \frac{\alpha+1}{\alpha}\theta_s \tag{1}$$

$$\theta_m^* = \theta_s - \frac{\Delta\theta}{\alpha} \tag{2}$$

$$\theta_f = \theta_s + \Delta\theta \tag{3}$$

$$T_f = \frac{T_m}{\alpha} = -\frac{T_s}{1+\alpha} \tag{4}$$

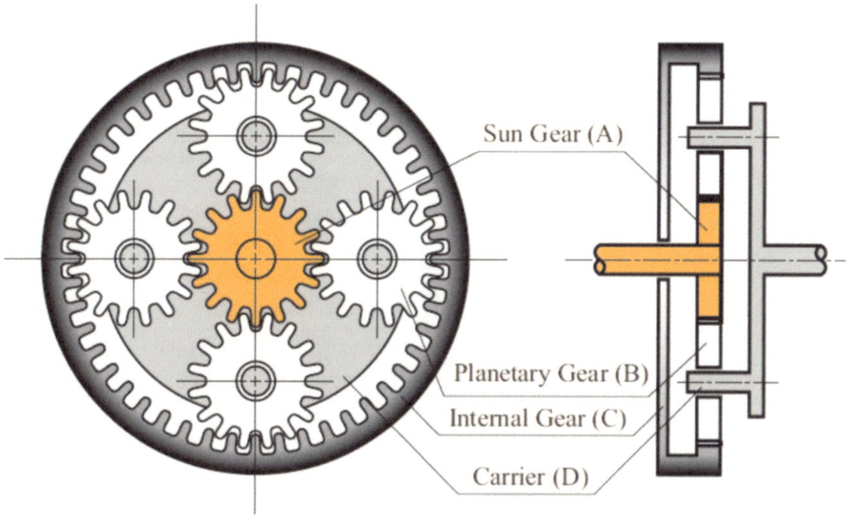

Fig. 1. A structure of planetary gear.

T_s	Steering wheel torque
T_d	Driver torque
T_r	Reaction torque to a driver
T_f	Front axle wheel torque
T_f^*	Front axle wheel reference torque
T_{RTRS}	Reaction torque from road surface
T_a	Power assist torque
θ_s	Steering wheel angle
θ_m	Ring gear angle
θ_m^*	Ring gear reference angle
θ_f	front axle wheel angle
θ_t	front wheel angle
$\Delta\theta$	Angle between steering wheel and front axle wheel
J_s	Steering wheel inertia
J_f	Front steering inertia
B_s	Friction coefficient of steering wheel
C_f	Friction coefficient of front steering
K_a	Power assist ratio
G_t	Gear ratio of rack and pinion

Table 1. Steering system parameters.

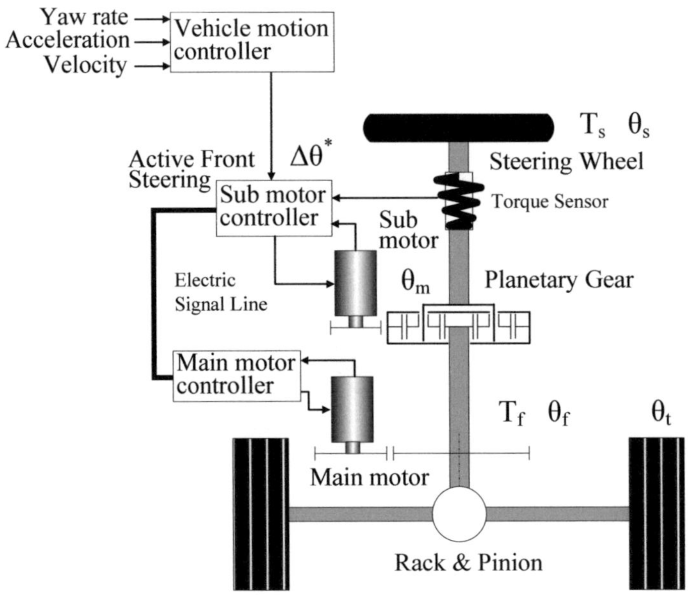

Fig. 2. Differential steering system.

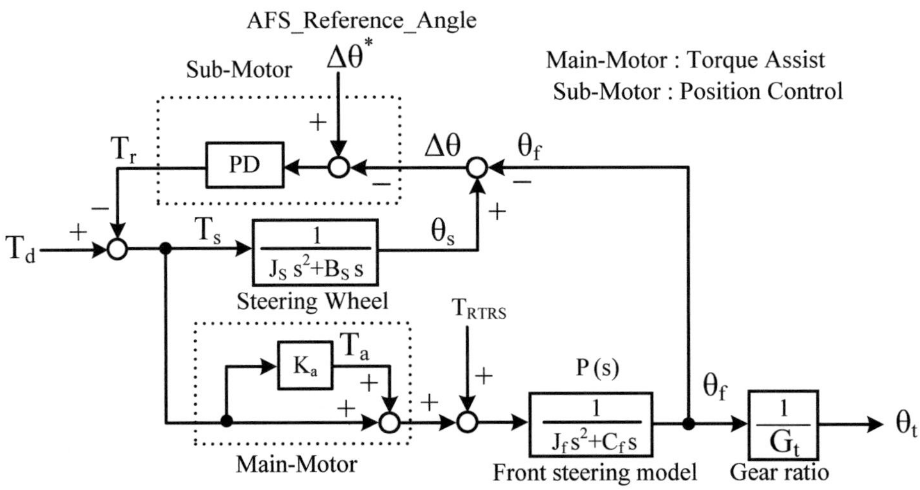

Fig. 3. Block diagram of active front steering control.

3. Method of experimental evaluation

We verified the following effects conclusively with a steering device shown in Fig.4.

Firstly, evaluation of the response of the following differential angle to the reference differential angle by AFS. Block diagram of AFS is shown in Fig.2. If a motion control PC that supervises vehicle's motion detects a dangerous movement such as a slip, the PC sends the differential angle reference to the PC controlling the steering system. The steering PC attempts to match the differential angle to the reference. This study evaluated the response.

Secondly, Evaluation for reducing the steering interference during AFS. This paper evaluated a way to reduce the steering interference that impedes driver steering during AFS operation.
Finally, Evaluation for transmitting the reaction torque required for safe driving. In order to realize driver safety, it is neccesary to provide an appropriate reaction torque for the steering wheel. This study evaluated a way to transmit the reaction torque to the driver via the steering wheel.

3.1 Steering device

The AFS simulation was evaluated using the steering device. The photo of planetary gear our produced is shown in Fig. 5 and courtesy photograph of steering device around the planetary gear is shown in Fig. 6. The block diagram shown in Fig. 2. The state variable used in the block diagram is shown in Table 1. This device consisted of two motors and a planetary gear. The sub-motor controlled the planetary gear so that the differential angle $\Delta\theta$ matched the reference angle $\Delta\theta^*$ for AFS. $\Delta\theta$ is differential angle between the steering wheel and the front axle wheel angle. The main-motor was used to reduce the steering load to help the driver. The motor also simulated the reaction torque T_{RTRS} that the road surface exerts on the front wheels. The equation for the reaction torque is shown in (5). This torque consists of self aligning torque (SAT) and the friction of the front wheels. It is transmitted in a direction opposite to that of the steering operation so that the driver does not turn the steering wheel excessively; thus, the torque is exerted to ensure safe driving. By using digital signal processor (DSP), this device can simulate AFS using SBW or a differential gear.

$$T_{RTRS} = K_t\theta_t + C_t\dot{\theta}_t \tag{5}$$

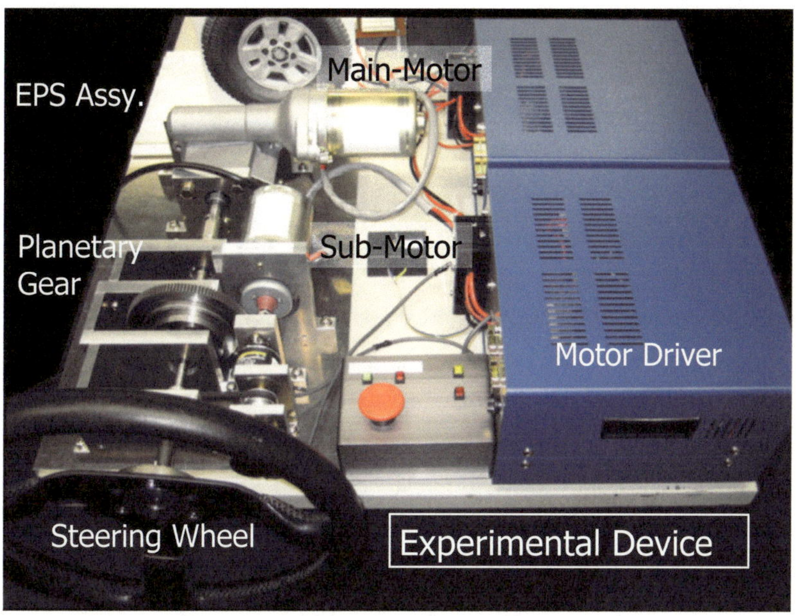

Fig. 4. Steering device.

Fig. 5. Planetary gear produced in Hori laboratory

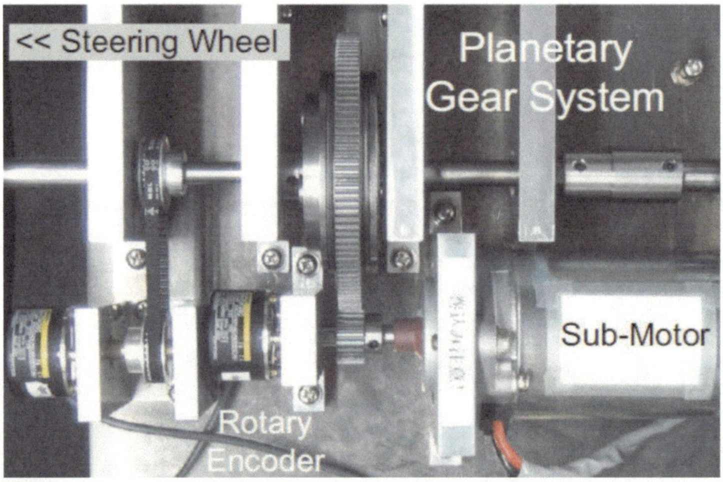

Fig. 6. Courtesy photograph of Steering device around the planetary gear

3.2 AFS simulation starting conditions

The experiment started when the steering wheel was turned to the left by 90°. After a few seconds, as soon as the motion control PC detected a slip in the vehicle, the AFS was activated. The AFS set the angle between the steering and front wheels to the reference angle $\Delta\theta^*$ calculated in the motion control PC. The experimental device simulated the reference angle, as shown in Fig. 7. For example, the AFS set the angle to -60° in the case of pattern 1, and passed on this value to the steering control PC. This means that the front wheel's angle was reduced by 30° in the steering control PC. Then, each angle is shown in Fig.8. In comparison with the angular velocity for pattern 1, that for pattern 5 was faster while that for pattern 9 was slower. The device could simulate 12 types of patterns in an AFS experiment.

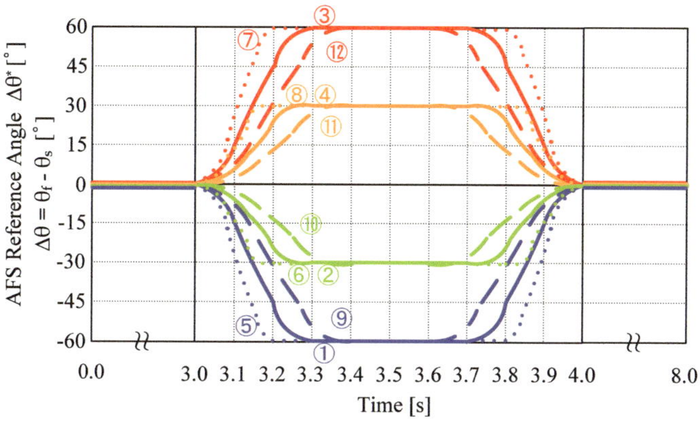

Fig. 7. Differential reference angle for AFS.

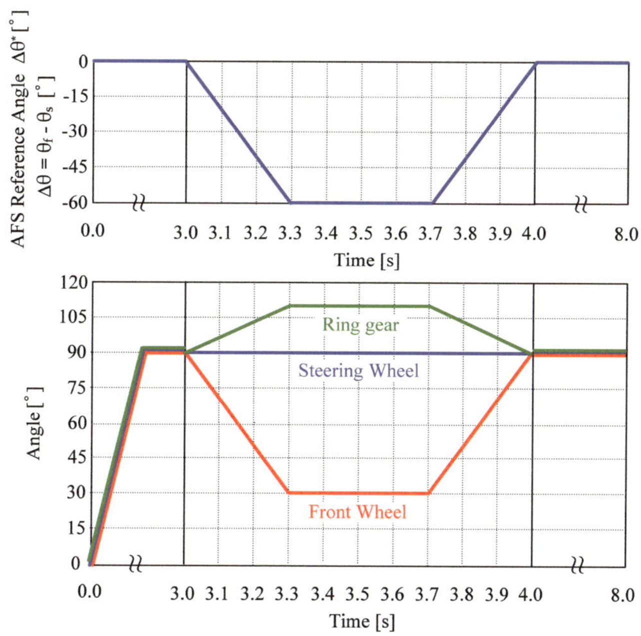

Fig. 8. Angle control for AFS.

4. Fundamental verification of driver sensitivity during afs operation

The verification revealed that driver steering sensitivity interfered with automatic steering control. Differential angle control was applied in the experiment. We performed the reference angle experiment using patterns 1 to 12, as shown in Fig. 7, and verified the effect of driver sensitivity. From the results, we proposed a reaction torque control method based on driver sensitivity.

4.1 Comparative verification of steering interference in differential angle

4.1.1 Pattern 1: reference angle Δθ*=-60(°)

The experimental result of the reference angle based on pattern 1 is shown in Fig. 9. The graph of the angle shows that the control could achieve the differential angle (-60°) in 3 to 4 s when AFS was performed, but there were two problems. One problem was that the steering wheel angle was considerably turned toward the other side of the front axle wheel angle. This prevented the driver from perceiving the vehicle's motion. Another problem was that the control caused the steering torque to decrease rapidly. This shows that the reaction torque decreased rapidly and it prevented the driver from operating the vehicle safely. That is to say, AFS returned the front axle wheel angle and the angle became small. The reaction torque that the driver received from the road surface decreased and the driver perceived that the steering wheel became light. The driver turned the steering wheel increasingly by 100°. This is a problem known as steering interference by AFS. If the driver maintained the steering wheel angle of 90°, the front axle wheel angle should have returned to the original 30°. However, in the results, it only returned to 40°. This means that sufficient AFS control was unrealizable from the viewpoint of motion control.

4.1.2 Pattern 2: reference angle Δθ*=-30(°)

The experimental result of the reference angle based on pattern 2 is shown in Fig. 10. The graph of the angle shows that the steering interference was less than the result of Fig. 9 since the front axle wheel angle was reduced to 60° and the steering angle was kept constant. As a result, the slope of the reaction torque that the driver received from the road surface was small. Therefore, it is advisable not to transmit the torque to the driver directly when the slope is excessively large. With regard to the steering axle, the reaction torque that would not cause steering interference is not more than 3 to 5 Nm/s.

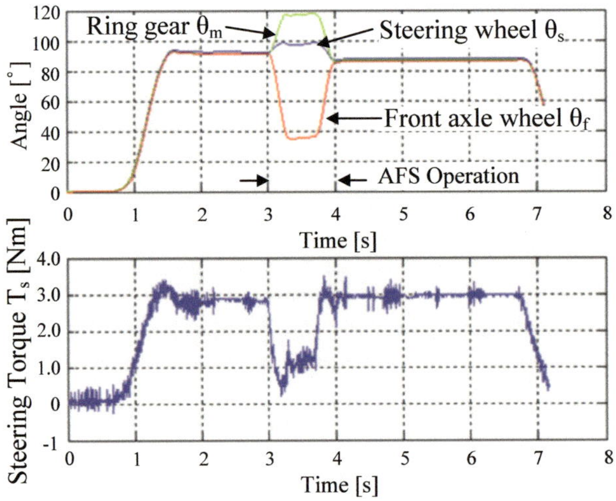

Fig. 9. Experimental result of reference angle based on pattern1.

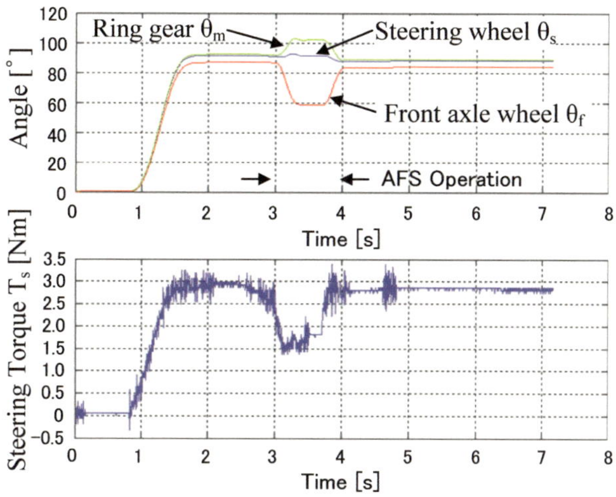

Fig. 10. Experimental result of reference angle based on pattern2.

4.1.3 Pattern 3: reference angle $\Delta\theta^*$=+60(°)

The experimental result of the reference angle based on pattern 3 is shown in Fig. 11. This result shows less steering interference than the result of Fig. 9. The result indicates that the driver sensitivity was insufficient to handle a reaction torque along a direction opposite to that of the torque that the driver received from the road surface before AFS operation.

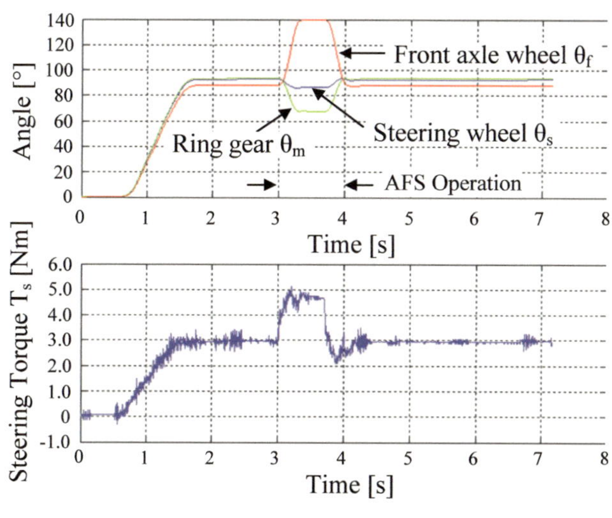

Fig. 11. Experimental result of reference angle based on pattern 3.

4.2 Comparative verification of steering interference in velocity of differential angle

The verification revealed that the velocity of the differential angle during AFS operation affected any steering interference caused by driver sensitivity.

4.2.1 Pattern 5: velocity of reference angle Δθ*/dt =-300(°/s)

The angular velocity for pattern 5 was higher than that for pattern 1. The experimental result for the reference angle based on pattern 5 is shown in Fig. 12.

4.2.2 Pattern 9: velocity of reference angle Δθ*/dt =-150(°/s)

The angular velocity for pattern 9 was slower than that for pattern 1. The experimental result of the reference angle based on pattern 9 is shown in Fig. 13.
A comparison of the experimental results for patterns 1, 5, and 9 revealed that a higher angular velocity in the front axle wheel led to greater steering interference. The driver received reaction torque in proportion to the velocity via the steering wheel. If the velocity was excessively high, the driver was unable to handle the torque and keep the steering wheel steady. Achieving safe AFS requires a torque to compensate for driver sensitivity to the velocity.

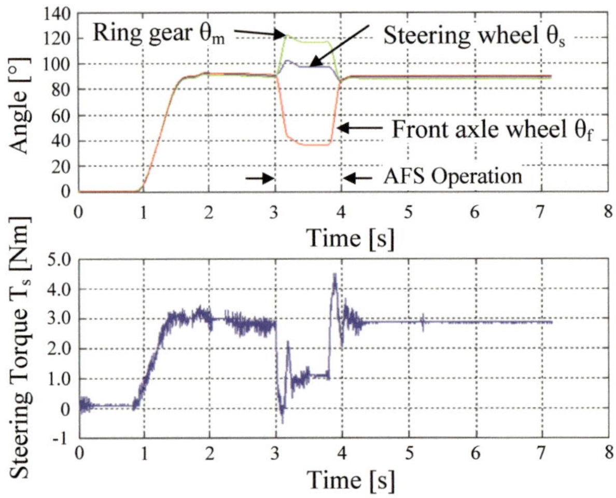

Fig. 12. Experimental result of reference angle based on pattern 5.

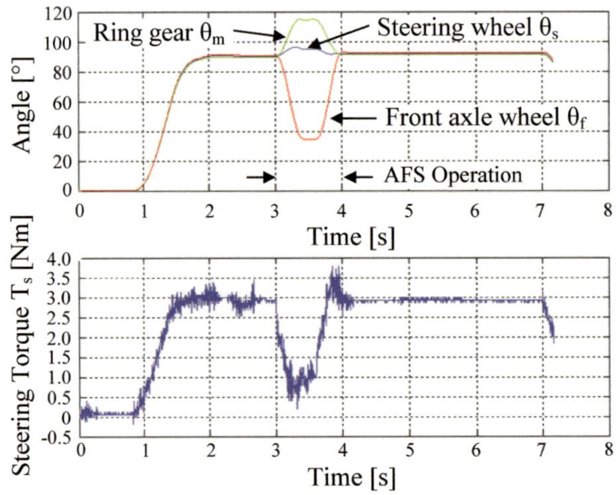

Fig. 13. Experimental result of reference angle based on pattern 9.

4.3 Reaction torque control based on idealized model with Steer-By-Wire

The block diagram of the control method is shown in Fig. 14. The control can decouple the steering interference completely by transmitting the reaction torque associated with the idealized model to the driver. Equation (6) for the reaction torque T_r shows that T_r is not associated in any way with the front wheels. The control can decouple the interference since the driver does not receive theAFS-modified reaction torque from the road surface. However, the driver loses the road information at the same time because there is no feedback from the front wheels to the steering wheel. Our control method realizes AFS by sending the reference angle of the front axle wheel to the front wheel controller, as shown in (7).

$$T_r = K_s\theta_s + C_s\dot{\theta} \tag{6}$$

$$\theta_f^* = \theta_s + \Delta\theta^* \tag{7}$$

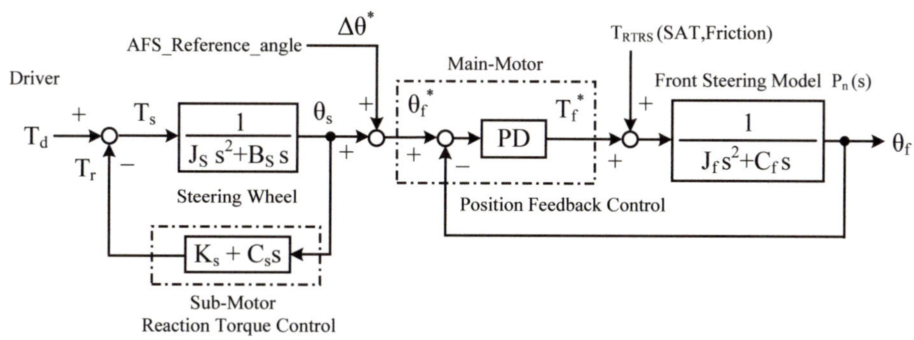

Fig 14. Block diagram of reaction torque control based on idealized model.

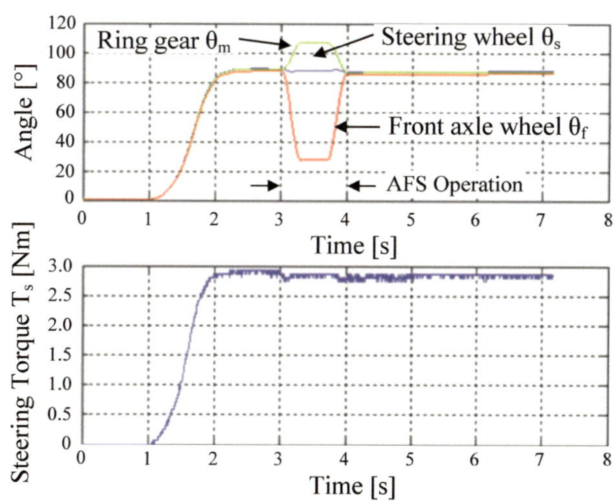

Fig. 15. Experimental result of AFS by reaction torque control based on idealized model with SBW.

The experimental results for this control are shown in Fig. 15. The graph of the angle shows that the front axle wheel angle returned to 30°, and the steering wheel angle was maintained constant angle at 90°. On the other hand, the steering torque shows that the driver constantly received a reaction torque. This control method did not cause any steering interference because no AFS-modified reaction torque was transmitted by the control from the road surface to the driver. Although this control method seems to be suitable, a problem encountered is that the driver operates the vehicle without receiving any reaction torque as road information when the front wheels hit a bump in the road or the driver operates the vehicle on a gravel road. In other words, the control can decouple the interference completely but at the same time it cannot provide reaction torque required for safe driving.

5. Reaction torque control based on driver sensitivity during afs operation

Since the differential steering always transmits the reaction torque from road surface to the driver, during AFS operation It led to steering interference. On the other hand, SBW could decouple the steering interference because no AFS-modified reaction torque was transmitted to the driver. However, it could not provide reaction torque as road information to the driver. Reducing steering interference and transmitting road reaction torque are inconsistency to each other. We propose two reaction torque control methods for a driver. First one is reaction torque control based on variable assist ratio control, another one is based on reaction torque observer and hysteresis torque control.

5.1 Reaction torque control based on variable assist ratio with differential steering

The block diagram of this method is shown in Fig. 3 and It controls assist ratio K_a based on differential angle between steering wheel and front axle wheel. This equation is shown in (8). K_{a0} is constant value. Reaction torque to a driver is shown in (9). Torque difference between steering angle and front axle wheel angle is transmitted to a driver. It is same as torque sensor with torsion bar spring for conventional EPS system. On the other hand, the equation of front axle wheel angle is shown in (10). It shows that the variable assist torque can comepensates the reaction torque T_{RTRS} from road surface. Because the reaction torque depends on the front axle wheel angle as SAT. To the next, it compensats the front axle wheel angle. As a result, It can control the reaction torque transmitted to a driver.

$$K_a = \left(\theta_f / \theta_s\right) K_{a0} \tag{8}$$

$$T_r = \left(\theta_s - \theta_f\right)\left(K_P + K_D s\right) \tag{9}$$

$$\theta_f = \frac{\left[(1+K_a)T_s + T_{RTRS}\right]}{J_f s^2 + C_f s} \tag{10}$$

5.2 Reaction torque control based on estimated reaction torque and driver sensitivity

As mentioned above, although the reaction torque control based on the idealized model with SBW can decouple the steering interference, a limitation of the control method is that the driver cannot use the reaction torque as road information. In an application based on this control method, the reaction torque must be measured or estimated to transmit this information to the driver. In this study, we estimated the torque using an assist motor attached to the front axle wheel and an observer. In addition, we propose to adjust steering sensitivity by transmitting model reaction torque to a driver. Block diagram of this method is shown in Fig. 16.

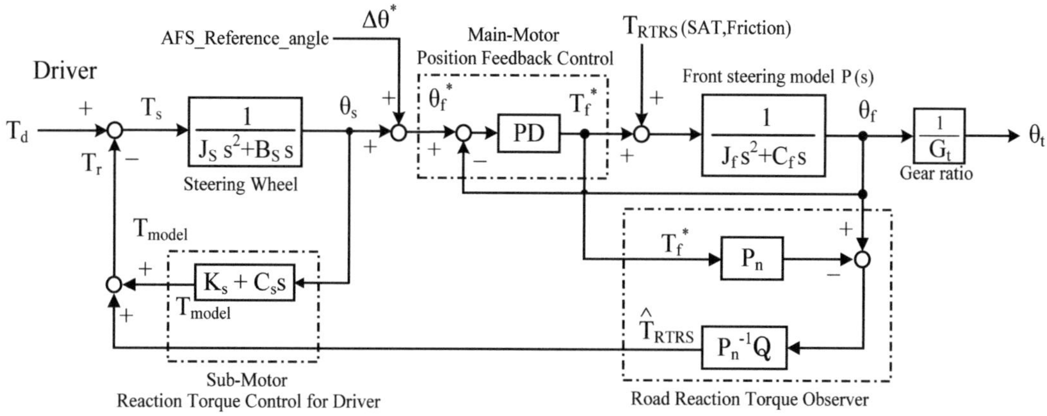

Fig. 16. Reaction torque control based on estimated reaction torque and driver sensitivity.

5.2.1 Reaction torque observer

The block diagram of the reaction torque observer is shown in Fig. 17. The reaction torque T_{RTRS} from the road surface is an unknown state variable to be estimated. First, the estimated angle of the front axle wheel, $\hat{\theta}_f$, is calculated using the nominal model $P_n(s)$ of plant $P_n(s)$ and the reference torque of the front axle wheel, T_f*, which are known state variables. Second, $\Delta\theta$ is calculated as the difference between the estimated value $\hat{\theta}_f$ and the measured value θ_f. $\hat{T}_{RTRS}$ is calculated by means of the inverse model P_n^{-1} of the plant. Since P_n^{-1} is not a proper function, $\hat{T}_{RTRS}$ is calculated using a low-pass filter Q. If the reaction torque $\hat{T}_{RTRS}$ estimated by the observer were to be directly transmitted to the driver, the control system would cause steering interference in the same manner as the conventional angle control, as shown in Fig. 9. Therefore, it is necessary to adjust the gain and frequency of $\hat{T}_{RTRS}$ with the low-pass filter Q. The formula is shown below.

$$\theta_f = \left(T_f^* + T_{RTRS}\right)P(s) \qquad \hat{\theta}_f = T_f^* \cdot P(s) \tag{11}$$

$$\hat{T}_{RTRS} = \left(\theta_f - \hat{\theta}_f\right)P_n^{-1}(s)\cdot Q(s) \tag{12}$$

If the nominal model $P_n(s)$ is confirmed to be identical to plant P(s), $\hat{T}_{RTRS}$ is calculated by the following formula (13).

$$\hat{T}_{RTRS} = T_{RTRS}(s)\cdot Q(s) \tag{13}$$

P(s) is identified using the Prediction Error Method based on the Maximum-Likelihood Method.

Low pass filter Q reduces higher frequency gain than cut-off frequency $1/\tau_q$. τ_q is time constant value. In this paper, cut-off frequency $1/\tau_q$.is 20 (Hz) because reaction torque required for steering operation is about 0-10 [Hz]. And feedback gain G_{fb} is adjusted by formula (15) to assist driver return the steering wheel. As a result, it is able to reduce the steering interference.

$$G(s) = \frac{G_{fb}}{1 + 2\tau_q s + \tau_q^2 s^2} \tag{14}$$

$$1 < G_{fb} < 1.5 \tag{15}$$

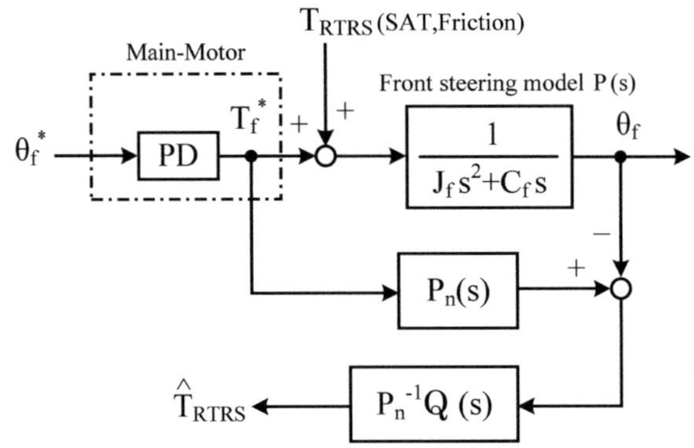

Fig. 17. Reaction torque observer.

5.2.2 Hysteresis torque control for varying steering sensitivity

This method compensates for driver sensitivity during the steering operation. To be specific, it controls hysteresis, which compensates for driver sensitivity. The sub-motor controls hysteresis by adjusting the coefficient C_s in proportion to the angular velocity of the steering wheel in Fig. 16. The model reaction torque shown in (16) controls steering sensitivity for driver. The coefficient C_s is determined by difference angle $\Delta\theta$ from the formula (17). When a driver operates the steering wheel angle of sine-wave at 0.5 [Hz], lissajous curve is drawn To evaluate the steering characteristics. This result is shown in Fig. 18. Blue line is K_s = 2.0, C_s = 0.2, red one is K_s = 2.0, C_s = 0.6. When the coefficient C_s is large, the hysteresis band is expanded. One the other hand, when comparing to frequency of steering torque in Fig.19, high frequency steering torque vibration is reduced. However, large hysteresis band strikes the driver as heavy steering feeling. For this reason, steering system need to adjust aproapriately hysteresis band according to road reaction torque. When the reaction torque is changed using AFS, an appropriate hysteresis value enables the driver to maintain the operation and reduces the vibration of the steering torque and the steering interference. Because driver has not felt the vibration in the hysteresis band. The torque T_r transmitted to the driver is given by (18). In this paper, the stiffness K_s is zero because of compensating only steering vibration.

$$T_{model} = K_s\theta_s + C_s\dot{\theta}_s \tag{16}$$

$$K_s = \frac{K_t}{1 + K_a}\ ,\ C_s = \frac{C_t}{1 + K_a}\frac{d(\Delta\theta)}{dt} \tag{17}$$

$$T_m = \hat{T}_{RTRS} + T_{model} \tag{18}$$

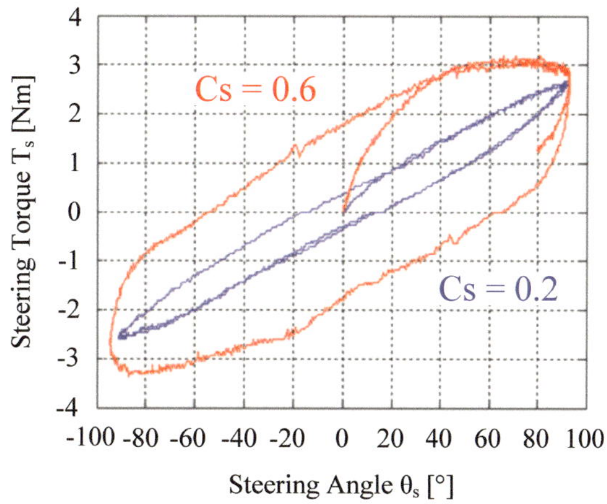

Fig. 18. Hysteresis band characteristic changed by viscous friction.

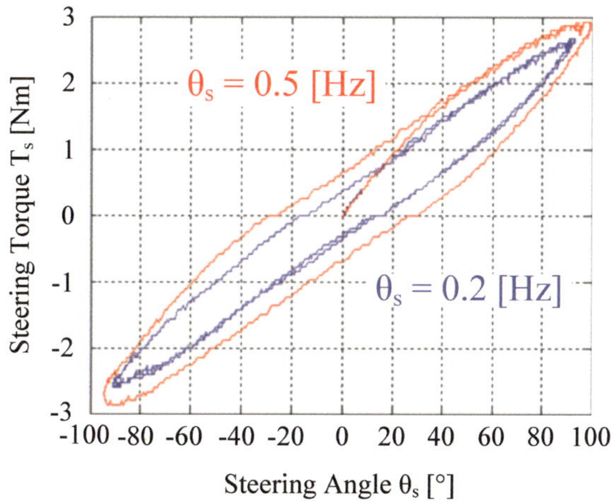

Fig. 19. Hysteresis band characteristic changed by angular velocity of steering wheel.

6. Experimental verification of AFS by proposed reaction torque control for steering

The experimental verification revealed that the proposed method was effective in allowing a driver to operate a vehicle safely with AFS. The reference angle of the AFS experiment was applied to patterns 1 and 5, which caused large steering interference at the experiment of 4th clause.

6.1 Reaction torque control based on variable assist ratio control

6.1.1 Pattern 1: reference angle $\Delta\theta^*$=-60(°)

The experimental result of the control is shown in Fig. 20. The steering torque shows that the reaction torque is more constant than the result of the differential angle control shown in Fig. 9. In the graph of the angle, the steering wheel angle has been kept constant angle at 90° and the front axle wheel angle was returned to 30°. As the two results, the control is an effective technique to reduce the steering interference during AFS operation.

6.1.2 Pattern 5: velocity of reference angle Δθ*/dt =-300(°/s)

The experimental result is shown in Fig. 21. The steering interference has been reduced compared to that of differential angle control shown in Fig. 12. The front axle wheel angle has been returned to about 30°. However, the steering torque has been vibrated in 3.0-3.3(s) and 3.7-4.0(s). When the front wheel angle is returned quickly, road reaction torque contains larger vibration. Since this method compensates reaction torque in proposion to the front wheel angle, the reaction torque vibration could not be reduced sufficiently.

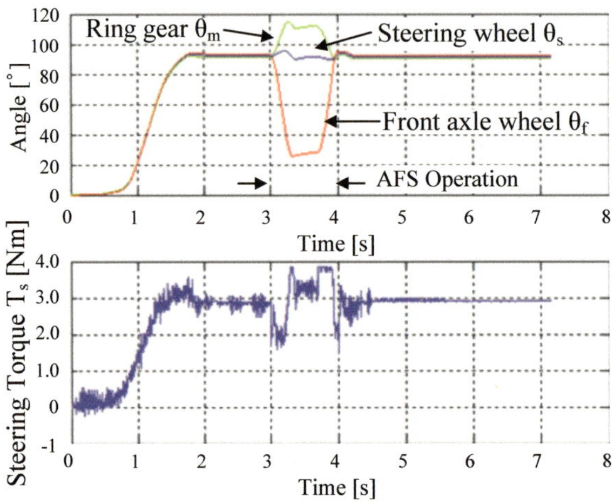

Fig. 20. Experimental result of AFS based on variable assist ratio control in cased of pattern 1.

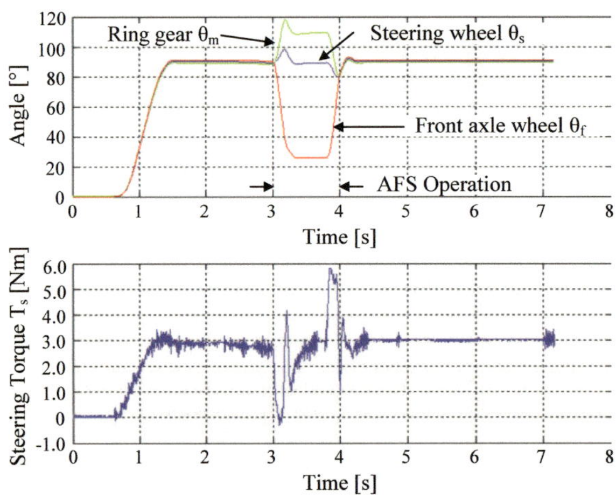

Fig. 21. Experimental result of AFS based on variable assist ratio control in cased of pattern 5.

6.2 Reaction torque control based on estimated reaction torque and driver sensitivity

6.2.1 Pattern 1: reference angle Δθ*=-60(°)

The experimental result is shown in Fig. 22. The graph of the angle shows that the technique did not cause the same steering interference as the reaction torque control based on the idealized model shown in Fig. 15; in addition, the proposed technique exhibited excellent performance in matching the reference angle of the front axle wheel. On the other hand, the graph of the steering torque shows that the technique could transmit a small amount of the reaction torque from the road surface to the driver.

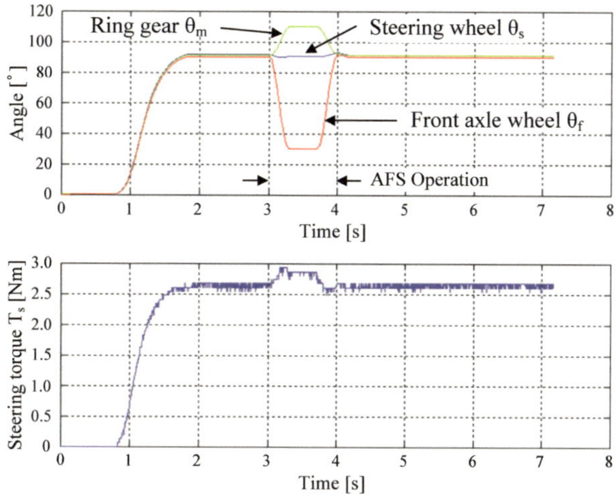

Fig. 22. Experimental result of AFS by reaction torque control based on driver sensitivity in cased of pattern 1.

Thus, the technique was able to decouple the steering interference while simultaneously transmitting information from the road surface. From the driver's viewpoint, it is essential to ensure that a small amount of torque is transmitted to the driver. If such transmission is realized, the driver can recognize AFS operation and sense the cooperation between the steering operation and the system assisting the driver's operation.

6.2.2 Pattern 5: velocity of reference angle Δθ*/dt =-300(°/s)

The experimental result is shown in Fig. 23. The use of our control method resulted in a large hysteresis value during AFS operation.

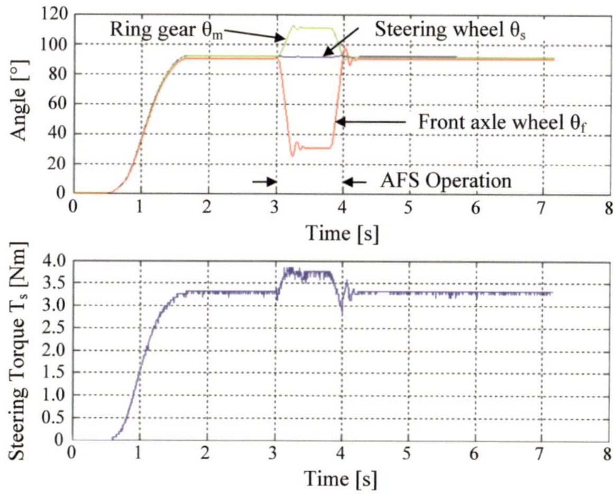

Fig. 23. Experimental result of AFS by reaction torque control based on driver sensitivity in cased of pattern 5.

The result showed that the control reduced the steering interference while simultaneously transmitting the reaction torque from the road surface to the driver. In addition, the control reduced the steering torque vibration by adjusting the hysteresis value.

7. Conclusion

In this study, we verified the relationship between steering interference associated with AFS-assisted automatic steering and driver steering sensitivity. Drivers are very sensitive to the reaction torque, and if a driver receives it directly after it has been changed by the AFS, safe operation is impossible. Moreover, the experimental results showed that driver sensitivity is nonlinear. In particular, drivers are unable to handle a reaction torque exerted along a direction opposite to that of the torque that the driver receives from the road surface before AFS operation. Therefore, we propose a reaction torque control method based on the driver sensitivity. To be specific, the proposed method controls the gain and frequency of the reaction torque from the road surface to prevent steering interference and allow the road information required for safe operation to be transmitted to the driver. In addition, it controls hysteresis to reduce steering torque vibration. As a result, the driver can operate the steering wheel safely with AFS.

8. References

Falcone, P et al. (2007). Predictive Active Steering Control for Autonomous Vehicle Systems . *IEEE Transaction. Control Systems. Technology*, Vol.15, No.3, pp.566-580, ISSN 10636536

Bianchi, N. et al. (2007). The Steering Effect PM Motor Drives for Automotive Systems. *IEEE Industry Applications Magazine*, Vol. 14, No. 2, pp. 40-48, ISSN 1077-2618

Endo, S. & Kobayashi, H. (2004). EPS Control Technology. *NSK Technical Journal Motion & Control*, No. 16, pp. 51-56, ISSN 1342-3630

Aoki, K. & Hori, Y. (2004). Realization of Active Steering with Steering System Using Planetary Gear. *IEE of Japan Industry Applications Society Conference*, 2-38, pp.425-428, September 2004

Nakano, S et al. (2000). Steering Control for Steer-by-Wire System. *KOYO Engineering Journal*, No. 158, pp.21-26, ISSN 0911-145X

Koehn, P & Eckrich, M. (2004). Active Steering The BMW Approach Towards Modern Steering Technology. *SAE Technical Paper Series (Society of Automotive Engineers)*, 01-1105, ISSN 0148-7191

Minaki, R & Hori, Y. (2009). Experimental Verification of Driver-Friendly Reactive Torque Control Based on Driver Sensitivity to Active Front Steering. *Industrial Electronics, 2009. IECON '09. 35th Annual Conference of IEEE*, pp. 3077-3082, ISSN 1553-572X

Minaki, et al. (2009). Ergonomic verification of reaction torque control based on driver's sensitivity characteristics for active front steering. *Vehicle Power and Propulsion Conference, 2009. VPPC '09. IEEE*, pp. 160-164, ISSN 1556-6072

Minaki, et al. (2008). Experimental Evaluation of Active Front Steering Control Based on Estimated Reaction Torque. *IEE of Japan Industry Applications Society Conference*, 2-17, pp.347-352, August 2008

Part 3

Effect on Efficiency of Motor with Torque Control

3

Study on the Energy Efficiency of Soft Starting of an Induction Motor with Torque Control

José de Oliveira[1], Ademir Nied[1],
Mário Henrique Farias Santos[1] and Rogério Pinho Dias[2]
[1]State University of Santa Catarina
[2]WEG Equipamentos Elétricos S.A.
Brazil

1. Introduction

Energy plays an important role in the paradigm of sustainable development. Nowadays, energy consumption is one of the key factors to measure the development of a country. According to the IEO2010 Reference case report (U.S. Energy Information Administration, 2010), world energy consumption will increase by 1.4 percent per year, from 495 quadrillion Btu in 2007 to 739 quadrillion Btu in 2035, as shown in Fig. 1.

The increasing of global energy consumption and the great concern to preserve the environment in the face of the current scenario which includes global warming, the greenhouse effect and the depletion of natural resources, aroused the awareness of governments, along with non-governmental organizations, to accelerate adopting policies and practices that will promote efficient use of energy (IEEE-USA, 2010).

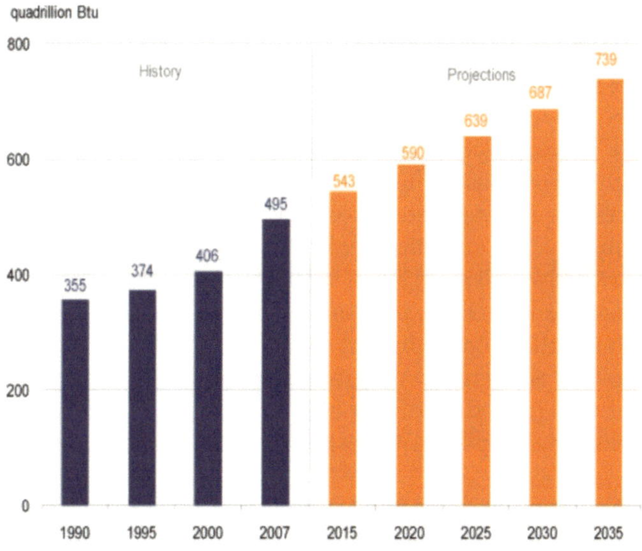

Fig. 1. World Marketed Energy Consumption, 1990-2035 (U.S. Energy Information Administration, 2010).

According to the WEC (World Energy Council, 2009), in industrialized countries three main sectors are responsible for three quarters of electricity consumption: electric motors, 45%, lighting, about 15%, and home appliances and consumer electronics, also around 15%. Global energy consumption of electric motors is around 9000 TWh per year. These data show the impact that electric motors produce on electrical energy consumption. Therefore, the manufacturers in this sector are investing heavily in the energy efficiency of their products.
From the establishment of the vector control approach for field-oriented control and the evolution of microelectronics and power electronics, the induction motor has come to occupy a prominent position among variable-speed electric drives (Leonhard, 1988).
The same technological developments that allowed the use of induction motors in variable speed drives to replace DC motors also enabled the development of soft starting electronic devices that allow better control of the induction motor starting and stopping process. This equipment, beyond controlling the starting current value suited to the demands of the energy utilities, allows better control of motor acceleration and deceleration (Bruce et al., 1984), (Colleran & Rogers, 1983), (Nevelsteen & Aragon, 1989).
In this chapter, the authors present a study of energy efficiency obtained by using a soft starter to start a 300 HP induction motor. Experimental results are shown comparing the energy efficiency of a soft starter using two different control techniques: current control and torque control.

2. Induction motor starting characteristics using a soft starter

Soft starters are devices that use silicon-controlled rectifiers (SCRs) for providing a low voltage to the motor from a grid voltage and fixed frequency. The value of the voltage applied to the motor can be varied over a wide range, from very small values up to the value of the grid by controlling the firing angle of the SCRs.
The consequences of this reduction in the value of the applied voltage are the decrease in electric current drawn from the grid and the reduction of the torque developed during the process of starting or stopping. Fig. 2 shows a basic schematic diagram of a soft starter with anti-parallel SCRs and the feedback from current and voltage.
The interface circuit is responsible for acquiring and processing measured voltage and current signals, which are processed digitally by the control program. The command signals from the digital processing are then sent by control to a firing circuit, responsible for driving the SCRs and consequently, the induction motor.
Soft start devices provide two major benefits in their application (Gritter et al., 2000):

- Less stress on the motor mechanically coupled to the load due to the reduced amplitude of the pulse of the starting torque motor;
- Smoothing the motor acceleration and reducing the demand for energy flow in electric power systems due to the reduction of peak current when starting the motor.

Usually two techniques are employed to control the soft starter: voltage-ramp and current control.
The voltage-ramp technique is performed through a firing-angle ramp of SCRs. Thus, although better than the non-electronic starter methods, it does not guarantee a more effective control over current and acceleration during starting process and over deceleration during stopping process.

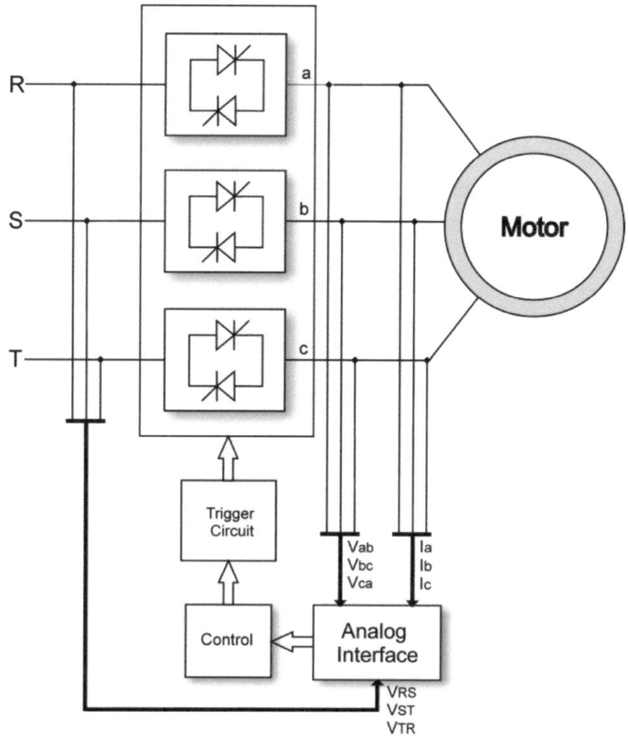

Fig. 2. Basic diagram of a soft-starter.

The angle ramp, shown in Fig. 3, is called voltage-ramp, due to the fact that the majority of commercial soft starters do not possess voltage feedback applied to the motor, i.e., the induction motor voltage is controlled through a firing angle ramp in an open loop. This technique is simple, and it is used in low-cost commercial soft starters. It will always produce a starting quadratic torque curve that can be applied to small hydraulic pumps and small fans.

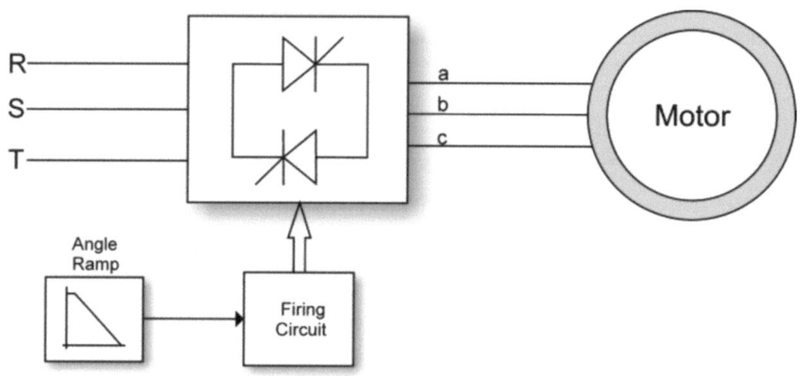

Fig. 3. Basic block diagram of the voltage-ramp technique.

Considering the starting motor using the voltage-ramp technique, the waveforms of electromagnetic torque and motor current are shown in Fig. 4.

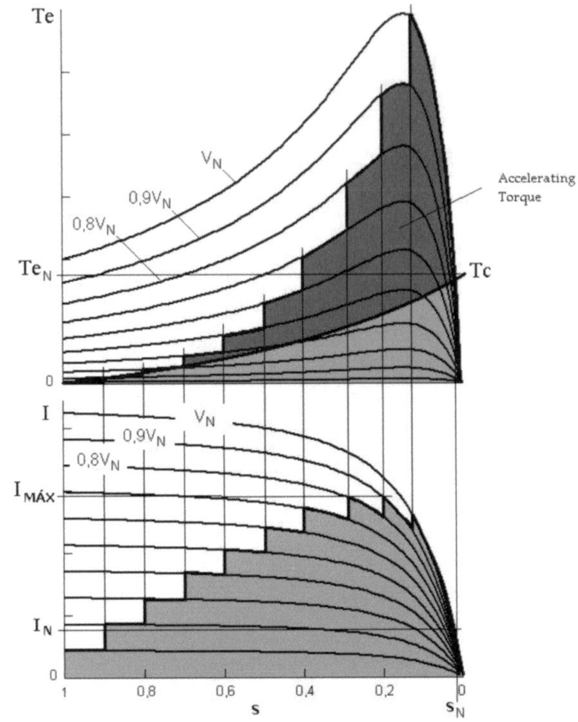

Fig. 4. Electromagnetic torque and motor current waveforms using the voltage-ramp technique.

Figure 4 illustrates the motor starting using the voltage-ramp technique with a load which has a quadratic torque curve (T_C). As can be verified, for each interval of 10% of synchronous motor speed, 10% of rated motor voltage is added. Still, the variation of electromagnetic torque (Te) and motor current (I) due to the variation of stator voltage can be clearly seen. Using this type of ramp-voltage technique, the electromagnetic torque produced by the motor is always quadratic (Palma, 1999).

On the other hand, the current control technique controls the current at a constant value and one that is appropriate for the requirements for starting, solving the problem of the high inrush current transient inherent in induction motors with direct starting. However, it does not satisfactorily solve the issue of stress on mechanical coupling with the load. It also can induce a more pronounced acceleration during the final period of the starting process (Zenginobuz et al., 2001).

Fig. 5 shows the diagram of the current control technique. The three current signals are used to calculate the true RMS current value. This current value (I_{RMS}) is used by the controller as the feedback signal in order to keep the current constant at the pre-set value (I_{REF}) during the motor starting process.

This technique will always generate an almost constant starting torque curve that can be applied to a constant load torque. This technique has the advantage of adapting the induction motor starting within the capabilities of the supplying utility.

For an example of a load with quadratic torque characteristics (Tc), Fig. 6 presents the induction motor electromagnetic torque and current waveforms considering the motor starting using the current control technique, i.e., with current limitation. The current is kept constant at a predefined value during the whole motor starting time, but the current control can only operate under a rated voltage variation of 10 %.

From Fig. 6, it can be noted that through the current limitation (I_{MAX}) the initial electromagnetic torque developed (Te) varies very little until 70% of the synchronous motor speed. However, during the final period of the motor starting process, one can verify the existence of a torque pulse, in which, for certain kinds of loads, an abrupt acceleration can be generated (Palma, 1999). This technique also has problems with the starting of non-constant loads, such quadratic loads, linear loads and loads with high resistive torque, because it is difficult to predict the torque profile produced in this type of technique.

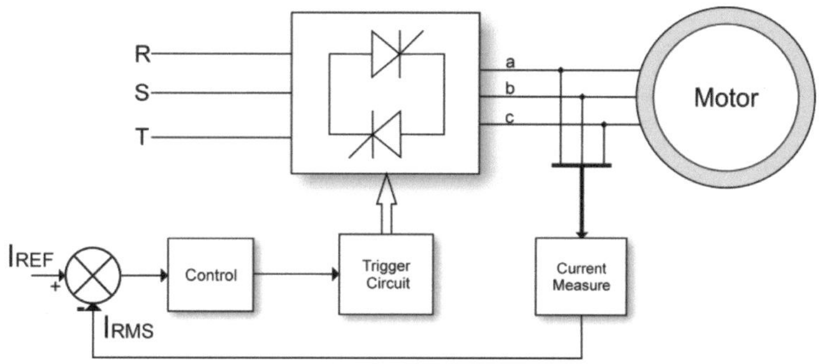

Fig. 5. Basic block diagram of the current control technique.

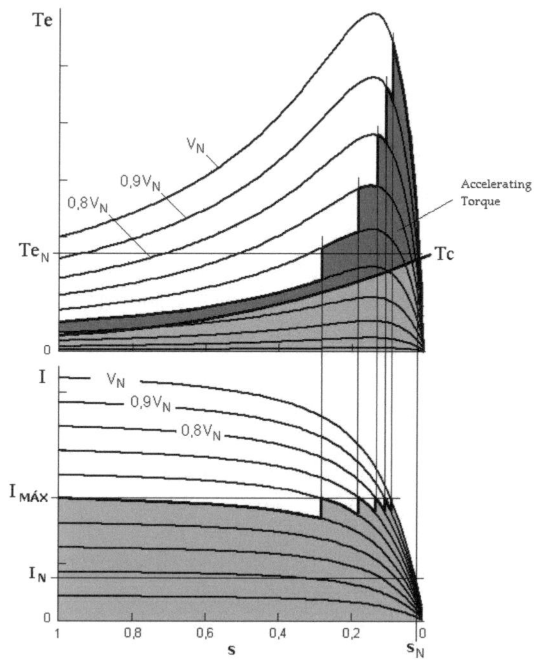

Fig. 6. Electromagnetic torque and motor current waveforms using the current control technique.

From the electromagnetic torque characteristics shown in Figs. 4 and 6, it is noted that, using the voltage-ramp technique or the current control technique for limiting current, in both cases a torque profile is obtained that does not adequately match with the load profile. Although these motor starting techniques can provide an almost constant torque during the beginning of the starting process, one can verify the existence of a torque pulse, in which for certain kinds of loads, an abrupt acceleration can be generated.

Therefore, it seems that a good solution to solve the existence of a torque pulse is to directly control the electromagnetic torque developed by the induction motor during the starting and stopping processes, adapting it to the torque profile of the driven load.

3. Torque control technique

The difficulties related to implementing a soft-starter with torque control are more related to limitations resulting from hardware (inability of the drive frequency variation due to SCRs) than the theoretical knowledge associated with the control design.

Figure 7 shows the diagram of the torque control technique for limiting the electromagnetic torque of the motor (Nied et al., 2008). From the motor voltages and currents measurements, the flux estimation is calculated, and from that, one obtains the estimated electromagnetic torque. The error signal generated from the difference between the torque reference signal and the estimated torque signal is applied to a controller, which generates the appropriate signal to the soft-starter trigger circuit.

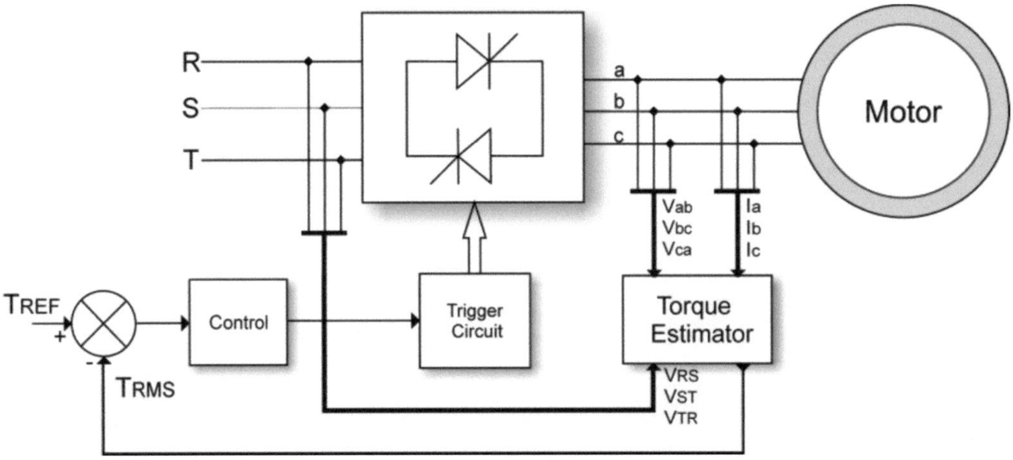

Fig. 7. Basic block diagram of the torque control technique.

In Equations (1-3), $\vec{E}$ is the back electromotive force, R_S is the motor stator resistance, $\vec{V}_S$ and $\vec{I}_S$ are, respectively, the measured stator supply voltages and currents in the $\alpha-\beta$ stationary reference frame:

$$\vec{E} = \vec{V}_S - \vec{I}_S.R_S \tag{1}$$

$$\vec{V}_S = \begin{bmatrix} V_{\alpha S} \\ V_{\beta S} \end{bmatrix} \tag{2}$$

$$\vec{I}_S = \begin{bmatrix} I_{\alpha S} \\ I_{\beta S} \end{bmatrix}. \tag{3}$$

The stator flux is estimated through the integration of the back electromotive force as described by the following equations:

$$\vec{\lambda}_S = \int \vec{V}_S - \vec{I}_S.R_S dt \tag{4}$$

$$\lambda_{\alpha S} = \int V_{\alpha S} - I_{\alpha S}.R_S dt \tag{5}$$

$$\lambda_{\beta S} = \int V_{\beta S} - I_{\beta S}.R_S dt. \tag{6}$$

From the previous equations, the flux magnitude and angle of the estimated stator flux can be written as:

$$|\lambda_S| = \sqrt{\lambda_{\alpha S}^2 + \lambda_{\beta S}^2} \tag{7}$$

$$\theta = \tan^{-1}\left(\frac{\lambda_{\beta S}}{\lambda_{\alpha S}}\right). \tag{8}$$

Finally, the electromagnetic torque of the motor can be estimated using the following cross product:

$$T_{em} = \frac{3}{2}.\frac{P}{2}.\left(\vec{\lambda}_S \times \vec{I}_S\right) \tag{9}$$

where P is the number of poles of the induction motor.

From Equation (4), one can note that the induction motor stator flux can be estimated through the integration of back electromotive force (Idris & Yatim, 2001), (Holtz, 2003), (Zerbo et al., 2005). This method requires only the stator resistance parameter. Besides, it is independent of the rotor position or the speed information. However, this method presents inherent problems related to pure integration, mainly the motor operating at nearly zero speed when the voltage drop in stator resistance become significant. This approach tends to be inaccurate because the integrators do not operate well at speeds near zero as there is no motor-induced back electromotive force.

When a pure integration is implemented in the discrete form, an error can arise. This error comprises the drift produced by the discrete integration and also the drift produced by measurement offset error present in the back electromotive force (Holtz, 2003), (Xu, X. & Novotny, D. W. 1991). A small DC component can drive the pure integration into saturation.

To reduce integration problems, (Nied et al., 2010) adopted a simple approach utilizing a low-pass filter to estimate the induction motor stator flux.

As can be verified in Fig. 8, using the estimated stator flux and the measurement stator current, the instantaneous electromagnetic torque produced by the induction motor is obtained. The flux estimation method is based on the following steps:

a. Using a high-pass filter after the stator voltage and current analog readings, designed with cutoff frequency one decade below the excitation frequency to eliminate DC offset presented in stator voltage and current analog readings, (Zerbo et al., 2005);
b. Using a low-pass filter designed with cutoff frequency one decade below the excitation frequency to substitute the pure integrator and to eliminate high frequency noises presented in analog readings (Idris & Yatim, 2001).

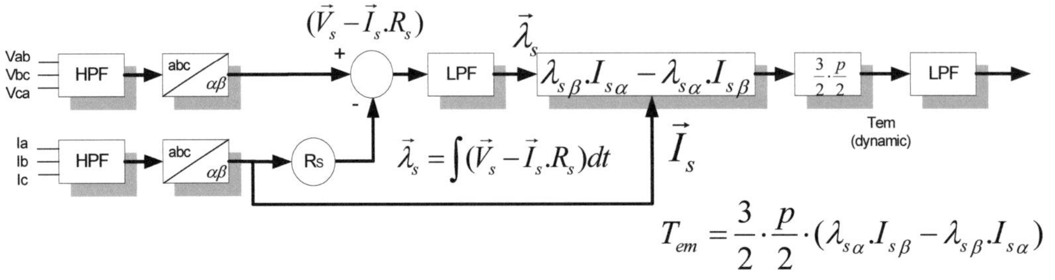

Fig. 8. Diagram of the induction motor electromagnetic torque estimator based on an integrator with offset minimization.

To eliminate the ripple presented in estimated electromagnetic torque, another first order low-pass filter was used. The second low-pass filter was designed in the same manner as the first low-pass filter, generating the filtered electromagnetic torque curve.
In the next section, the comparative results of two techniques are presented: current control and torque control. Using the torque control technique for tracking a reference torque results in an improved energy efficiency of induction motor starting.

4. Study of energy efficiency

Experiments were carried out to verify the performance of the torque control technique. One standard induction motor with the following motor data in physical units was tested: power 300 CV (220 kW), rated voltage 440 V, rated current 345 A, rated speed 1790 rpm, 4 poles. Due to the high current and power values, the experimental setup comprised a soft starter with torque control and a rated current of 365 A, a dc machine (300 kW) working as a dynamometer, an oscilloscope and an energy analyzer. Fig. 9 shows the hardware that was used to obtain the experimental results. In addition, it must be noted that, for the two first experiments the load torque (T_L) was a constant value.
The induction motor starting waveforms of speed, acceleration and RMS line current *versus* time are shown in Figs. 10–12. Three experiments were performed:

- Experiment 1: starting performance using the *current* control technique (Fig. 10);
- Experiment 2: starting performance using the *torque* control technique (Fig. 11);
- Experiment 3: starting performance using the *torque* control technique and considering *three reference points for* T_{REF} (Fig. 12).

Figures 10 and 11 show the starting results obtained from the motor using two different soft-starting strategies. As can be seen in Fig. 10, despite the fact that the line current is limited, there is an acceleration pulse, i.e., the acceleration is almost constant except at the final period of the starting process when the acceleration increases, at least, by two times. This behavior is not desired, and it may damage the mechanical system components.

Considering the current control technique, when starting the motor, the SCRs' trigger angle is controlled until the current limit is reached and remains under this limit until the motor reaches its rated speed, when the SCRs' trigger angle is minimum, i.e., the same as that of the supply voltage. The current can thus be kept constant at a predefined value during the whole motor starting time. However, with regard to the torque at the motor shaft, the existence of a torque pulse close to nominal rotation is due to the SCRs' trigger angle which becomes minimum, imposing a feeding voltage and sudden motor acceleration.

Fig. 9. View of the hardware used to obtain experimental results.

In Fig. 10, it can be noted that the motor starting time is about 25 seconds. It can be verified too that the current maximum value was limited to around two times as the nominal current value.
It can be observed that the current control technique limits the high inrush current transient inherent in induction motors with electromagnetic line starters. Besides, it can be noted during most of the motor starting process that the acceleration is constant and around 1 rot/s². However, during the final period of the motor starting process, one can observe an abrupt acceleration around 2.5 times more than the constant acceleration.

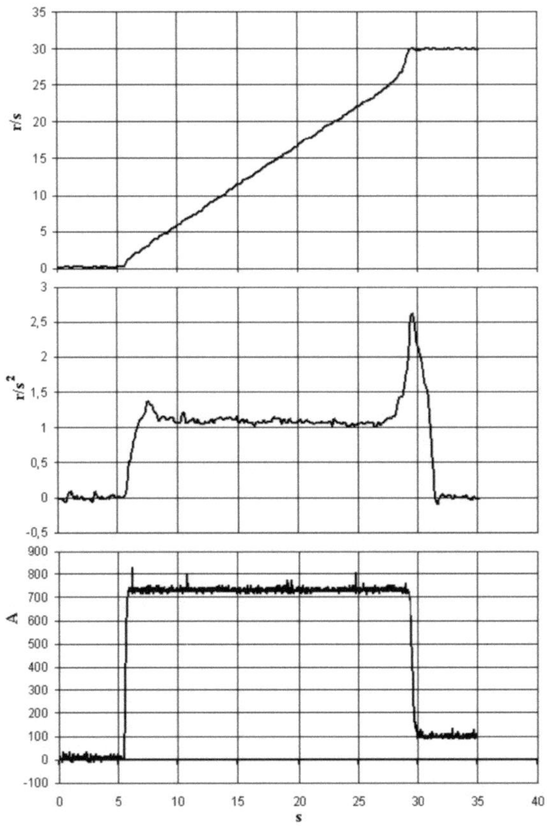

Fig. 10. Speed, acceleration and current obtained in Experiment 1 using the current control technique.

Measures of energy consumption for this experiment were taken. The measured value for the energy consumed due to active power was 0.531 kWh and the amount due to reactive power was 3.592 kVArh.
Similar current behavior is verified for the torque control technique in Fig. 11, which can be obtained by setting T_{REF} to a constant value, i.e., one point of reference. However, in this case, there is no torque pulse, the speed is quite smooth, and an abrupt acceleration is no longer recognized. Observe that the control is acting directly on the torque profile which is different from the current-control case, where the control did not act directly on the torque profile.

From the torque control technique characteristics, it can be verified that:

- The speed slope is linear throughout the starting process, avoiding the pulse acceleration recorded in the current control technique;
- The acceleration waveform presents approximately linear behavior;
- The current waveform was very close to that observed in the current control technique.

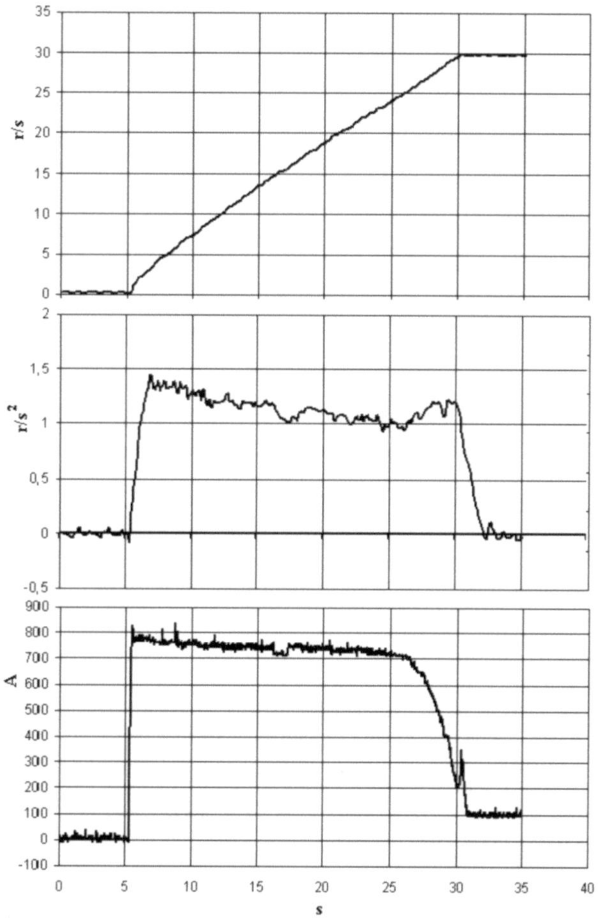

Fig. 11. Speed, acceleration and current obtained in Experiment 2 using the torque control technique.

Measures of energy consumption for Experiment 2 were taken too. The measured value for the energy consumed due to active power was 0.550 kWh and the amount due to reactive power was 3.691 kVArh.

In Fig. 12, three reference points for T_{REF} were considered. In this case, unlike the waveforms shown in Fig. 11, a torque ramp of two different shapes was tailored: In the first-half starting period, a rising torque ramp was defined, and after that, a descending torque ramp was defined. This behavior can be observed in the acceleration waveform. As a consequence, the speed waveform looks like an S with a smooth behavior at the beginning and at the end of the starting period. The motor RMS line current follows the shaft-torque reference behavior.

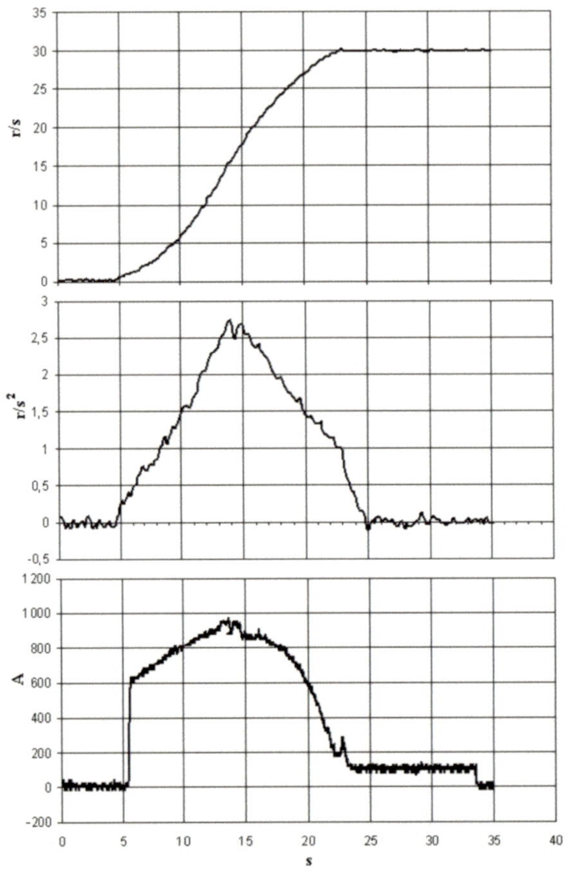

Fig. 12. Speed, acceleration and current obtained in Experiment 3 using the torque control technique and considering three reference points for T_{REF}.

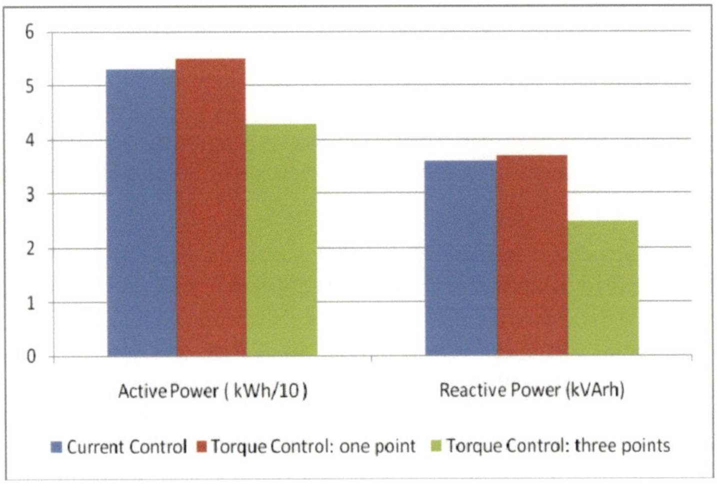

Fig. 13. Bar graph of the experimental results.

Measures of energy consumption for Experiment 3 were taken. The measured value for the energy consumed due to active power was 0.430 kWh and the amount due to reactive power was 2.484 kVArh.
The bar graph shown in Figure 13 compares the values of consumed energy (active power and reactive power) for the three experiments.
Comparing the values of energy consumed in Experiments *1* and *2*, it appears that the difference is not appreciable. This fact can be justified because of the similarity in the behavior of the magnitudes shown in Figs. 10 and 11.
Based on the values of energy consumption obtained in Experiments *1* and *3*, there is an energy savings of about 19% in active power and reactive power at 31%. These savings of energy were expected due to the fact that in Experiment 3, a current value higher than in Experiments 1 and 2 was allowed in a shorter time.

5. Conclusion

Minimum standards for energy efficiency have been the basis for policies used by states and government to save energy. The study presented in this chapter followed the idea of developing innovations that bring better performance. It was shown that with a new control technique for the soft-starter, it is possible to save energy by improving induction motor drive.
The soft-starters for working with fixed frequency are devices that have some limitations for the implementation of more sophisticated control strategies, for example, those employed in three-phase inverters used in controlled speed drives. Despite this fact, it was shown that it is feasible to implement a soft-starter with control of electromagnetic torque. Using this control technique in the soft-starter allows the adjustment of motor torque curve to the needs of the driven load, thus improving the performance of the drive during transient starting and stopping. This technique becomes important in cases where the motor is subjected to a regime of very intense work, such as elevators and large commercial presses.
The experimental results showed the effectiveness of the presented technique, eliminating the peak torque occurring at the end of the motor starting process when using the current control technique. Moreover, depending on the flexibility to adapt to the type of load inertia, made possible by adjusting the torque reference, it was shown that it is possible to obtain an energy savings of about 19% in active power and reactive power at 31% depending on the type of drive that you need to perform.

6. Acknowledgments

The authors wish to acknowledge the support provided by UDESC and WEG Equipamentos Elétricos for this investigation.

7. References

Bruce, F. M.; Graefe, R. J.; Lutz, A. & Panlener, M. D. (1984). Reduced-Voltage Starting of Squirrel-Cage Induction Motors. *IEEE Transactions on Industry Applications*, Vol.IA-20, No.1, (January/February 1984), pp. 46-55, ISSN 0093-9994.
Colleran, P. J. & Rogers, W. E. (1983). Controlled Starting of AC Induction Motors. *IEEE Transactions on Industry Applications*, Vol.IA-19, No.6, (November/December 1983), pp. 1014-1018, ISSN 0093-9994.

Gritter, D. ; Wang, D. & Habetler, T. G. (2000). Soft Starter inside Delta Motor Modeling and Its Control, *Proceedings of Industry Applications Conference*, ISBN 0-7803-6401-5, Rome, Italy, October 2000.

Holtz, J. (2003). Drift-and Parameter-Compensated Flux Estimator for Persistent Zero-Stator-Frequency Operation of Sensorless-Controlled Induction Motors. *IEEE Transactions on Industry Applications*, vol.39, No.4, (July/August 2003), pp. 1052-1060, ISSN 0093-9994.

Idris, N. R. N. & Yatim, A. H. M. (2001). Direct Torque Control of Induction Machines with Constant Switching Frequency and Improved Stator Flux Estimation, *Proceedings of IEEE 27th Annual Industrial Electronics Society Conference*, ISBN 0-7803-7108-9, Colorado, USA, November/December 2001.

IEEE-USA. (19.11.2010). Energy Efficiency, In: *Energy & Environment - National Energy Policy Recommendations*, 23.03.2011, Available from: <www.ieeeusa.org/policy/positions>.

Leonhard, W. (1988). Field-orientation for controlling AC machines - principle and application. *Third International Conference on Power Electronics and Variable-Speed Drives*, ISBN 0-85296-364-5, London, UK, July 1988.

Nevelsteen, J. & Aragon, H. (1989). Starting of Large Motors – Methods and Economics. *IEEE Transactions on Industry Applications*, Vol.25, No.6, (November/December 1989), pp. 1012-1018, ISSN 0093-9994.

Nied, A.; Dias, R. P.; de Oliveira, J.; Campos, R. de F & Marques, L. C. de S. (2008). Soft Starting of Induction Motor with Torque Control, *Proceedings of Industry Applications Society Annual Meeting*, ISBN 978-1-4244-2278-4, Edmonton, Canada, October 2008.

Nied, A.; de Oliveira, J.; de Farias Campos, R.; Dias, R. P. (2010). Soft Starting of Induction Motor With Torque Control. *IEEE Transactions on Industry Applications*, Vol.46, No.3, (May/June 2010), pp. 1002-1010, ISSN 0093-9994.

Palma, J. C. P. (1999). *Accionamentos Electromecânicos de Velocidade Variável*, Fundação Calouste Gulbenkian, ISBN 972-31-0839-9, Lisboa, Portugal.

U.S. Energy Information Administration. (27.07.2010). World Energy Demand and Economic Outlook, In: *International Energy Outlook 2010*, 23.03.2011, Available from: <www.eia.doe.gov/oiaf/ieo/world.html>

World Energy Council. (06.2009). Energy Efficiency Can Lead the Economic Recovery, In: *WEC Comment*, 23.03.2011, Available from: < www.worldenergy.org/news__events/wec_comment/2118.asp>

Xu, X. & Novotny, D. W. (1991). Implementation of Direct Stator Flux Orientation Control on a Versatile DSP Based System. *IEEE Transactions on Industry Applications*, Vol.27, No.4, (July/August 1991), pp.694-700, ISSN 0093-9994.

Zenginobuz, G.; Cadirci, I.; Ermis, M. & Barlak, C. (2001). Soft-Starting of Large Induction Motors at Constant Current with Minimized Starting Torque Pulsations. *IEEE Transactions on Industry Applications*, Vol.37, No.5, (September/October 2001), pp. 1334-1347, ISSN 0093-9994.

Zerbo, M.; Sicard, P.; Ba-Razzouk, A. (2005). Accurate Adaptive Integration Algorithms for Induction Machine Drive over a Wide Speed Range, *IEEE International Conference on Electric Machines and Drives*, pp. 1082-1088, ISBN 0-7803-8987-5, San Antonio, USA, May 2005.

Part 4

Sensorless Torque Control

4

Sensorless Torque/Force Control

Islam S. M. Khalil and Asif Sabanovic
Sabanci University
Turkey

1. Introduction

Motion control systems represent a main subsystem for majority of processing systems that can be found in the industrial sector. These systems are concerned with the actuation of all devices in the manufacturing process such as machines, robots, conveyor systems and pick and place mechanisms such that they satisfy certain motion requirements, e.g., the pre specified reference trajectories are followed along with delivering the proper force or torque to the point of interest at which the process occurs. In general, the aim of force/torque control is to impose the desired force on the environment even if the environment has dynamical motion.
The previous motion requirements can be achieved when torque and position/velocity sensors are utilized. However, it is commonly agreed that force/torque sensors have many drawbacks such as (Iida & Ohnishi, 2003; Katsura et al., 2007):

- Sensor noise: the force/torque sensing process depends on amplifying the strain gauge strain. Therefore, the noise signal is amplified along with the measured signal.
- Narrow bandwidth: due to the previous sensor noise problem, force measurement has to be realized through a low-pass filter. Therefore, the bandwidth of the force sensor is limited by the bandwidth of the sensor noise.
- Complicity and non-collocation: utilization of force/torque sensors add an extra degree of freedom to the control system due to the soft structure of the force sensor, i.e., an energy storage element will exist between the actuated degree of freedom and the end effector in contact with the environment
- Instability: the previous problems cause instability
- The complex electronic setups and their associated wirings with each force or torque sensor embedded to the control system.

The previous drawbacks make the force measurement process undesirable and motivated many researchers over the past few decades to develop force observers to provide alternatives for the problematic force sensing process (Katsura et al., 2007). It was also shown that stable contact between the robot end effector and environment in the presence of force sensing is difficult due to the non-collocation and the soft structure of the force sensor (Li & Chen, 1998). The bandwidth of the force control using force observers is analyzed and the tradeoffs between stability and response were studied by many researchers to provide a comprehensive guidelines for force observer utilization (Katsura et al., 2008). In addition, the sensorless

force/torque control techniques found many applications in variety of disciplines such as bilateral teleoperation, microsystems and micro manipulation. Furthermore, the previous force observers can be combined with most of the relevant existing force control techniques such as the hybrid position/force control, stiffness control, the resolved acceleration based force control and the impedance control to provide an efficient tool which can be used in most of the industrial force control applications. The sensorless force control techniques are not only useful in the sense of avoiding force sensors but also in reducing the complicity of the entire control system and increasing their stability. In addition, this work present a novel sensorless force servoing techniques based on the action reaction state observer (Khalil & Sabanovic, 2011) which allows realizing the sensorless force control without taking any measurement from the robot or the environment except a single measurement form the robot actuator. The conventional force observer requires having velocity measurement of the end-effector in contact with the environment and using it as basis for the estimation process. The action reaction state observer allows realizing the interaction force observer without measuring position or velocity of the non-collocated end-effector in contact with the environment (Khalil & Sabanovic, n.d.), therefore allowing the force control to be realized with minimum amount of measurements from the dynamical system.
This chapter is organized as follows: Modeling of the force sensing process is included in Section 2 in order to illustrate the effect of the force sensing on the force control process. Force observers are analyzed and included in Section 3 which can be conceptually considered as alternatives for force/torque sensors. In section 4, a novel force control framework is presented which allows realization of the force control without force sensors and even with less measurements than those needed to construct the conventional force observers, thus reducing effect of the noise induced by each embedded sensor to the system. Experimental results are conducted on two experimental setups, a microsystems workstation and a single input lumped mass spring system with multiple degrees of freedom in contact with environment with varying impedance. The experimental results are distributed throughout the paper in order to provide a graphical interpretation and to support the presented ideas. Eventually, Conclusions and final remarks are included in Section 5.

2. Modeling of force/torque sensing

Contact between a robot end-effector with single degree of freedom and an environment in the presence of force sensing is illustrated in Fig. 1 where m and m_s are the masses of the

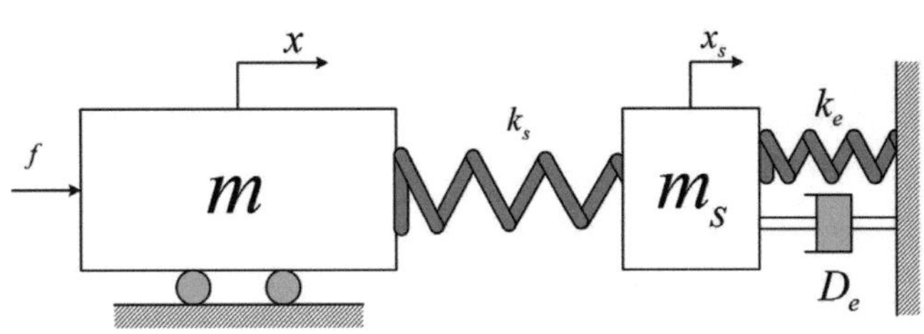

Fig. 1. Force sensor model

robot end-effector and the force sensor respectively. k_s, k_e and D_e are force sensor stiffness, environmental stiffness and the environment viscous damping coefficient, respectively. x and x_s are the positions of the robot end-effector and the force sensor, respectively. The following motion equations describe the contact mechanism depicted in Fig. 1

$$m\ddot{x} + k_s(x - x_s) + f_{dist_1} = f \tag{1}$$

$$m_s\ddot{x}_s - k_s(x - x_s) + f_{dist_2} = -f_{ext} \tag{2}$$

$$f^{ext} = z_e x_s \tag{3}$$

where z_e is the environmental impedance. f_{dist_1}, f_{dist_2} and f^{ext} are the disturbance forces on the first and second system degrees of freedom and the interaction force with the environment, respectively. In a regular force servoing problem the control system is designed such that the robot end-effector exerts a force f_{ext} that is equal to the desired force reference f^{des}. It is worth noting that in the previous example the environment is assumed stationary therefore its acceleration is not included among the previous motion equations. In addition, the environmental impedance can be modeled with either an energy storage element and energy dissipation elements or both. Therefore, in the presence of a force sensor, an extra degree of freedom is added to the system. In addition, the previous model shows that the end-effector is non-collocated via an energy storage element with stiffness k_s due to the soft structure of the force/torque sensor. The root locus of the system when the force sensor is attached to the robot end effector is depicted in Fig. 2-b, whereas, Fig. 2-a illustrates the root locus of the same system in the absence of force sensor. In both cases the root locus is plotted for different values of the environmental stiffness gain k_e. It can be easily shown from Fig. 2 that adding a force sensor to the dynamical model of the system affects the system stability dramatically. Controlling a collocated point is much easier then a non-collocated point along the dynamical system as it can be shown from Fig. 2. In order to impose the desired force on

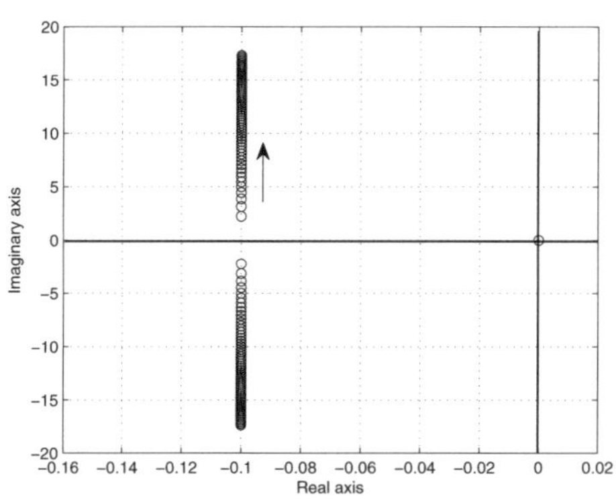

(a) collocated system

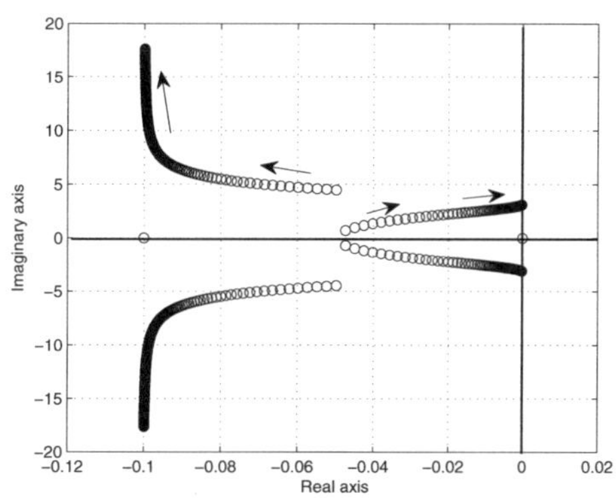

(b) non-collocated system

Fig. 2. root locus of system with and without force sensor for $k_e = 0 \longrightarrow 300$

the environment, the force measurement taken through a force sensor has to be used to realize the motion control system as depicted in Fig. 3. However, due to the force measurement nature at which the strain is measured and amplified along with the noise v^s, the actual interaction force with the environment cannot be precisely measured. In Fig. 3, k_v and k_p are the velocity feedback gain and proportional gains, respectively (Katsura & Ohnishi, 2007). k_r is the reaction force feedback gain. Due to the presence of noise in the force measurement, the force response of the force control system can be expressed as follows

$$\frac{f_{ref}^{ext}(s)}{f^{ref}(s)} = \frac{k_p k_f (k_e + sD_e)}{s(ms + k_v)(m_s s^2 + k_s + z_e) + k_s k_r (ms^2 + z_e) + k_p k_s (k_e + sD_e)} \quad (4)$$

$$\frac{f_n^{ext}(s)}{V^s(s)} = \frac{k_p k_f (k_e + sD_e)}{s(ms + k_v)(m_s s^2 + k_s + z_e) + k_s k_r (ms^2 + z_e) - k_p k_s (k_e + sD_e)} \quad (5)$$

The previous transfer functions describe the force response for both the reference and force sensor noise inputs, respectively. Therefore, the total response of the system can be written as the superposition of the reference input response $f_{ref}^{ext}(s)$ and the noise input response $f_n^{ext}(s)$

$$f^{ext} = f_{ref}^{ext}(s) + f_n^{ext}(s) \quad (6)$$

The previous equations indicate the effect of the force sensor noise on the desired force response. In addition, the force control parameters (k_v) and (k_p) have to be selected such that the characteristic equation of the transfer function (4) is stable. This can be easily done be using the denominator of (4) to formulate the Routh Hurwitz array and determine the gains such that stability is achieved according to the Routh Hurwitz stability criterion. The root locus of the force servoing system described by (4) is illustrated in Fig. 4 for different values of environmental stiffness (k_e) and velocity feedback gain (k_v). In general, environmental stiffness and the velocity feedback gain are of great importance for any force control problem since the cause of instability is that the environmental stiffness matrix increases the equivalent position gain of the control system. Therefore, the velocity gain has to be chosen not only based on the position gain but also on the large environmental stiffness. Otherwise the resulting system will be highly under damped and possibly unstable (Karunakar & Goldenberg, 1988). The previous analysis indicates how the utilization of

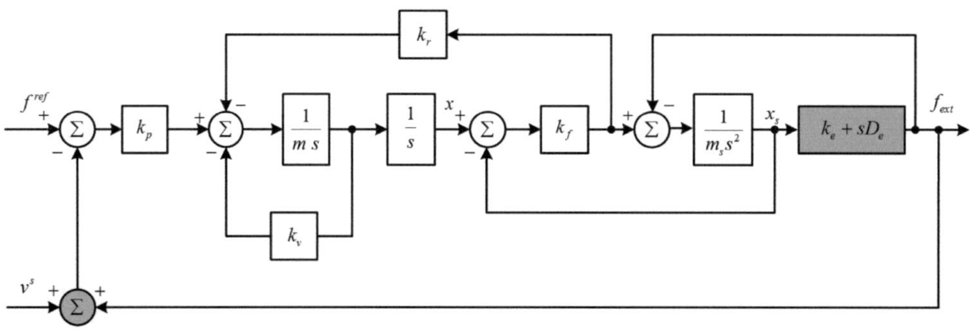

Fig. 3. Robot in contact with environment with force sensing

force sensors adds an extra degree of freedom to the control system, this in turn makes the robot end-effector non-collocated from its actuator. In addition, the response of the desired interaction force is affected by the noise input given by (5). Therefore, modeling of the force sensing shows how problematic the force sensing is, due to the non-collocation, the extra degree of freedom and the measurement noise. Before going into how to provide the sensorless force control techniques, we have to consider a more general form of the problem,

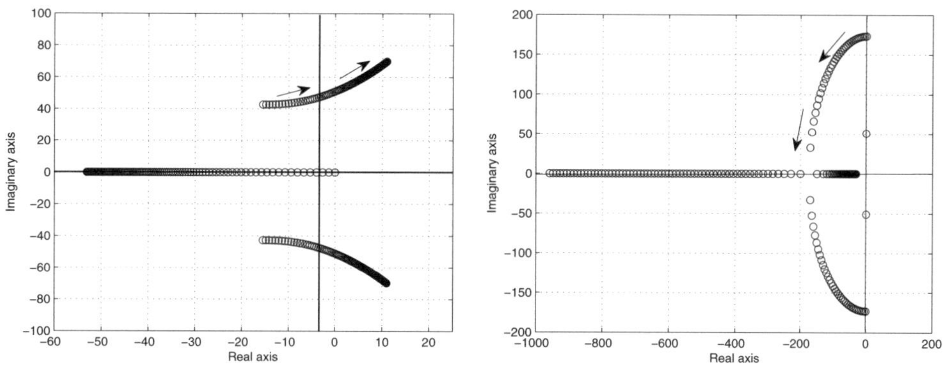

(a) $k_r = 1 \longrightarrow 10000$ (b) $k_v = 1 \longrightarrow 2000$

Fig. 4. root locus of system with force servoing

the generalized equation of motion of a robot manipulator can be therefore written as,

$$\mathrm{M}(q)\ddot{q} + \mathrm{V}\dot{q} + C(q,\dot{q}) + G(q) + \mathrm{J}^t f^{ext} = \tau \tag{7}$$

$$m_s\ddot{x}_s + k_s(x_e - x_s) - z_e x_s = 0 \tag{8}$$

where, q, $\dot{q}$ and $\ddot{q}$ are the joint position, velocity and acceleration vectors respectively. τ is the generalized joint torque vector. $\mathrm{M}(q)$ is the inertia matrix. V is the viscous damping matrix, $C(q,\dot{q})$ and $G(q)$ are the centrifugal and coriolis forces and gravity vectors, respectively. J is the robot Jacobian and f^{ext} is the interaction force with the environment. x_e is the robot end effector position. It will be shown in the upcoming sections that force observers and the other sensorless force control techniques will allow simplifying the dynamical model as the extra degree of freedom can be avoided.

3. Force/torque observers

In order to utilize force observers instead of the force sensors, we will first start with analyzing the disturbance observer structure since the reaction force observer is depending on the disturbance observer techniques. The conventional structure of the disturbance observer is depicted in Fig. 5. As depicted in Fig. 5, the reference input (u) along with the output (y) of the actuator can be used to estimate the incident disturbances (d). $\mathrm{G}(s)$ represents the model of the actuator while $(\mathrm{G}_n(s))$ represents the nominal model of the actuator. Therefore, disturbance can be computed as follows with force or torque units

$$\begin{aligned} d_o(s) &= u(s) - \mathrm{G}_n^{-1}(s)(y(s) - v(s)) \\ &= \left(\mathrm{G}^{-1}(s) - \mathrm{G}_n^{-1}(s)\right)y(s) + d(s) + \mathrm{G}_n^{-1}(s)v(s) \\ &= \Delta\mathrm{G}(s)y(s) + d(s) + \mathrm{G}_n^{-1}(s)v(s) \end{aligned} \tag{9}$$

where, $\Delta\mathrm{G}(s)$ is the real plant perturbation from the nominal one. Due to the presence of derivatives in the inverse plant dynamics $(\mathrm{G}^{-1}(s))$, the disturbance force has to be estimated

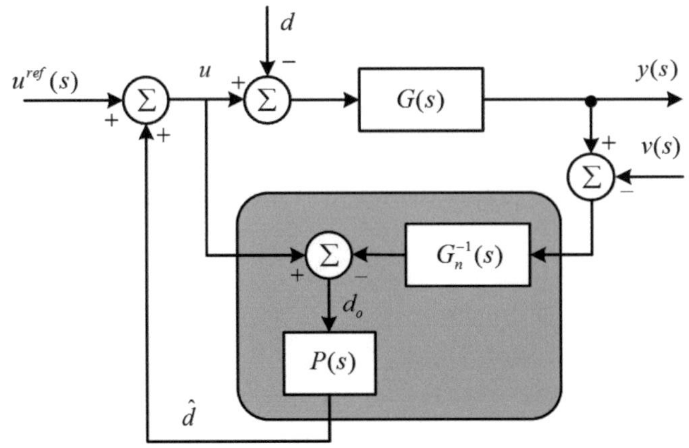

Fig. 5. Disturbance observer

through a low-pass filer as follows

$$\widehat{d}(s) = \mathrm{P}(s)d_o(s) \tag{10}$$

where (P(s)) is the low-pass filter associated with the disturbance observer. In general, disturbance can be modeled by

$$\frac{d^k d_o(t)}{dt^k} = 0 \tag{11}$$

and upon the degree of the polynomial used to model the disturbance signal, the order of the low-pass filter is determined. k is an integer that indicates the model of the disturbance (step, ramp, parabolic, etc) and can be further used in the selection of P(s). This in turn determines the order of the disturbance observer. Regardless to the order of the disturbance observer and the type of the disturbance signal, (9) along with (10) indicate that the estimated disturbance signal consists of the incident reaction disturbance force $d(s)$ along with parameter deviation induced disturbances $\Delta\mathrm{G}(s)y(s)$ and disturbance due to the sensor noise $\mathrm{G}_n^{-1}(s)v(s)$. Therefore, when the robot's end effector comes into contact with an environment, the interaction force is coupled in the disturbance signal. Therefore, reaction force observer depends on the disturbance observer structure expect that the other terms of the disturbance have to be subtracted in order to decouple the interaction forces from the disturbance signal. For instance, disturbance observer can be applied on (7) in order to estimate the disturbances, in this case disturbance force includes a coupled information which consists of the viscous damping friction, centrifugal and coriolis forces friction forces, gravity forces and external forces along with the parameter deviation induced disturbances, namely, the self-varied inertia and the torque ripple. Therefore, reaction force observer requires performing parameter identification in order to decouple the reaction force from the disturbance signal. According to (7), disturbance can be expressed as follows

$$d = \mathrm{V}\dot{q} + \mathrm{C}(q,\dot{q}) + \mathrm{G}(q) + \mathrm{J}^t f^{ext} + \Delta\mathrm{M}(q)\ddot{q} - \Delta k_t I_a^{ref} \tag{12}$$

where $\Delta\mathrm{M}(q)\ddot{q}$ and $\Delta k_t I_a^{ref}$ are the actuator self-varied inertia and torque ripple. Disturbance observer is generally implemented on the joint space of the robot at which an actuator is

located. Therefore, the last two terms of the disturbance force are related to the actuator parameters. k_t and I_a^{ref} are the actuator torque constant and reference current respectively. It is worth noting that equation (12) can be implemented on each joint of the robot. As shown in Fig. 5, the estimated disturbance with force or torque units is commonly used in the realization of an additional compensation control input which is used in the attainment of robustness.
In order to decouple the reaction force out of the disturbance force, or in order to construct a force observer for the realization of the sensorless force control, system parameters have to be identified. The constant angular velocity motion test (Murakami et al., 1993) can be used in order to estimate the gravity and friction forces at each degree of freedom of the system. A parameter identification process was proposed in (Khalil & Sabanovic, 2010a), in order to observe the self-varied inertia and the torque ripple online (Ohnishi et al., 1996). These parameter identification procedures have to be performed and used to assist in constructing the force observer. The estimated reaction force can be expressed as,

$$\widehat{f}^{ext} = \frac{g_{reac}}{s + g_{reac}} \left(I_a^{ref} k_{tn} + g_{reac} m_n \dot{q} - V\dot{q} - C(q, \dot{q}) - G(q) \right) - g_{reac} m_n \dot{q} \tag{13}$$

m_n and k_{tn} are the nominal actuator inertia/mass and torque/force constant. The interaction force with the environment is observed through a low-pass filter with a cut-off frequency g_{reac}. It can be easily concluded from (13) that the force observer depends on a single measurement from the actuator along with the reference current input. Therefore, the problems associated with force sensors can be avoided by utilization of the force observer (13). There exist two main points that we have to emphasize here about force observers, namely, the force observer sensitivity and the tradeoff between the response and the noise attenuation. Concerning with the second point, it can be easily shown that the reaction force is estimated through a low-pass filter in order to assist in attenuating the effect of noise amplification due to differentiation. Therefore, the bandwidth of the force observer or even the disturbance observer is limited with the bandwidth of the sensor noise. On the other hand, sensitivity of the reaction force observer is related to the accuracy of the position or velocity measurement. For instance, Fig. 6 indicates the force observer sensitivity for micro scale interaction forces. The observed force peak shown in Fig. 6 was obtained experimentally through a force observer of the form (13) based on position sensor with 10nm accuracy. This indicates that sensitivity of the disturbance or reaction force observer depends on the precision of the velocity or position measurement.
The Block diagram representation of the force servoing control system with force observer is depicted in Fig. 7. Force sensor is altered with the force observer, therefore the noise associated with the force sensor, the non-collocation and the complicity of the associated electronic setups and wirings are all avoided. Although the bandwidth of the control system can be relatively increased when force observer is utilized, still the noise added due to the the velocity measurement defines the new bandwidth of the force observer based force servoing system. In the experimental result illustrated in Fig. 6, the force observer requires the velocity information which was obtained by differentiating the position encoder signal through a low-pass filter with cut-off frequency g_l to attenuate the effect of noise amplifications due to differentiation, experimental values of the force observer gain and the low-pass filter gains are included in table. 1.
So Far, the previous analysis indicates that it is possible to alter force sensing with force observer. However, the main drawback in this procedure is that the bandwidth of the force observer based force servoing control system depends on the bandwidth of the noise

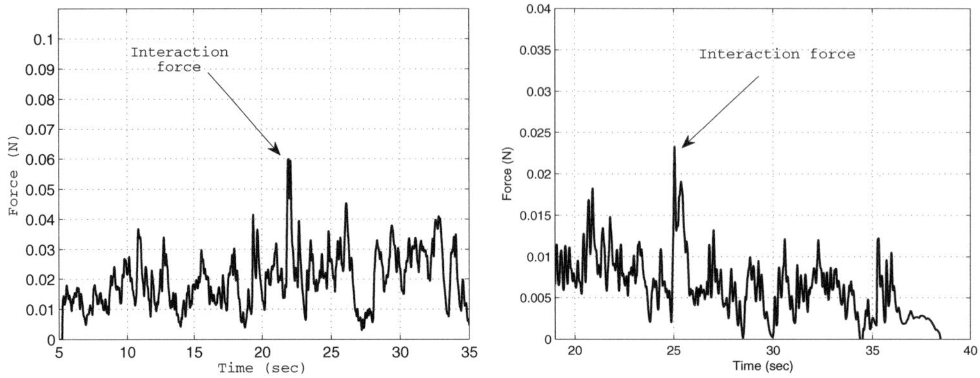

(a) interaction force (b) interaction force

Fig. 6. Sensitivity of reaction force observer for micro scale interaction force

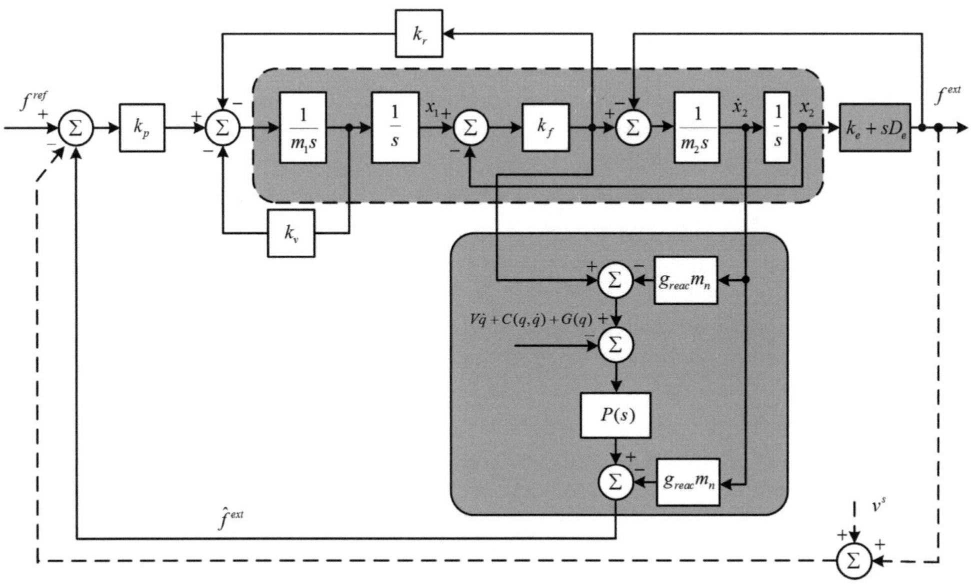

Fig. 7. Force servoing using reaction force observer

associated with the measured velocity signal. Although the force observer spares the control system from having a force measurement, it necessitates having a velocity measurement as shown in Fig. 7. This means that the force observer alters the force observer with a position or velocity sensor. Indeed, the noise associated with the position or velocity sensors is less than the noise associated with force sensor. Nevertheless, one would like to spare the control system having position or velocity sensor attached to the robot end-effector. With the observer presented in (Khalil & Sabanovic, 2010b;b), the position and velocity of the non-collocated end-effector can be observed from measurements taken from the input (actuator) of the dynamical system. This in turn, spares the non-collocated end-effector of the robot having any attached sensors.

In the next Section, a novel observer is introduced based on the action reaction law of dynamics in order to allow realization of the force observer without taking the velocity measurement from the dynamical system but rather its estimate.

4. Action reaction force control

It was shown in the previous Section that the force control can be realized without force sensor utilization through force observers. Therefore, the word sensorless force control does mean that force sensors is not utilized but it does not mean that sensors are not utilized since one has to measure or sense some signal from the system and use it as basis for the estimation process. The force observer explained in the previous section depends on measurement taken from the dynamical system which interacts with the environment, i.e., the velocity of the subsystem in contact with the environment. In this section, we will further try to avoid taking this measurement ($\dot{x}_2$) and/or (x_2).
It was shown in (O'Connor, 2003; 2007b), that sensorless control of flexible systems can be realized if the propagating mechanical waves throughout any dynamical system were considered as natural feedback (O'Connor, 1998; 2007a). This wave based approach depends on launching and absorbing mechanical waves with the right amount at the right time in order to position a non-collocated end-effector to a target position without residual vibration. However, the wave based approach requires having an additional measurement from a non-collocated mass along the flexible lumped system in order to determine the propagating mechanical waves along the system, in this section we would like to find a method to avoid this measurement. In (Khalil & Sabanovic, 2010b) the action reaction well-known law of dynamics was utilized in order to provide an instantaneous feedback-like force from a class of dynamical system with inaccessible outputs. The proposed action reaction state observer in (Khalil & Sabanovic, 2010b) allows realization of the previous force observer without measuring $\dot{x}_2$ or x_2. The action reaction state observer can be implemented for a class of dynamical system of the following form

$$\begin{aligned} \dot{\mathrm{x}} &= \mathrm{A}\mathrm{x} + \mathrm{B}u \\ \mathrm{y} &= \mathrm{C}\mathrm{x} + \mathrm{D}u \end{aligned} \tag{14}$$

which can be shown to have the following structure

$$\dot{\mathrm{x}}_a = \mathrm{A}_a \mathrm{x}_a + \mathrm{B}_a u_a + \mathrm{B}_{reac} f_{reac}(\mathrm{x}_\mathrm{a}, \dot{\mathrm{x}}_a, \mathrm{x}_\mathrm{p}^1, \dot{\mathrm{x}}_p^1) \tag{15}$$

$$\dot{\mathrm{x}}_p = \mathrm{A}_p \mathrm{x}_p + \mathrm{B}_p f_{reac}(\mathrm{x}_\mathrm{a}, \dot{\mathrm{x}}_a, \mathrm{x}_p^1, \dot{\mathrm{x}}_p^1) \tag{16}$$

where $\mathrm{x} \in \mathrm{R}^n$ and $\mathrm{y} \in \mathrm{R}^m$ are the state and measurement vectors, respectively. A, B, C and D are the system matrix, distribution vector of input, observation column vector and feed forward matrix with the appropriate dimensions, respectively. The dynamical system we consider consists of two subsystems, the first subsystem is the single input while the second is the flexible subsystem with the subscripts (a) and (p), respectively. x_a and x_p are the actuator and plant state vectors, respectively. Actuator states can be measured, whereas, the flexible dynamical subsystem states are inaccessible. $f_{reac}(\mathrm{x}_\mathrm{a}, \dot{\mathrm{x}}_a, \mathrm{x}_\mathrm{p}^1, \dot{\mathrm{x}}_p^1)$ is the incident instantaneous reaction force on the actuator. x_p^1 and $\dot{\mathrm{x}}_p^1$ are the position and velocity of the first non-collocated degree of freedom along the flexible subsystem. B_{reac} is the reaction force distribution vector. B_a is the actuator input distribution vector and B_p is the plant reaction force distribution vector. The action reaction state observer consists of the following three observers

$$\hat{d} = \frac{g_{dist}}{s+g_{dist}}[g_{dist}m_{an}\dot{x}_a + i_a k_{fn}] - g_{dist}m_{an}\dot{x}_a = \frac{g_{dist}}{s+g_{dist}}[i_a k_{fn} - s m_{an}\dot{x}_a] = \frac{g_{dist}}{s+g_{dist}} d \quad (17)$$

$$\widehat{f_{reac}}(i_a,\dot{x}_a) = \frac{g_{reac}}{s+g_{reac}}\left[g_{reac}\widehat{\Delta m_a}\dot{x}_a + i_a\widehat{\Delta k_f} + \hat{d}\right] - g_{reac}\widehat{\Delta m_a}\dot{x}_a \quad (18)$$

$$\dot{\hat{x}} = A\hat{x} + Bu + M\left(\widehat{f_{reac}}(x_a,\dot{x}_a,x_p^1,\dot{x}_p^1) - f_{reac}(x_a,\dot{x}_a,\hat{x}_p^1,\dot{\hat{x}}_p^1)\right) \quad (19)$$

Structure of the action reaction state observer is depicted in Fig. 8, this observer can be utilized in order to estimate the dynamical states of a subsystem at which measurements cannot be made. $g_{dist} \in R^+$, $g_{reac} \in R^+$ and $M \in R^{n\times 1}$ are the action reaction state observer scaler

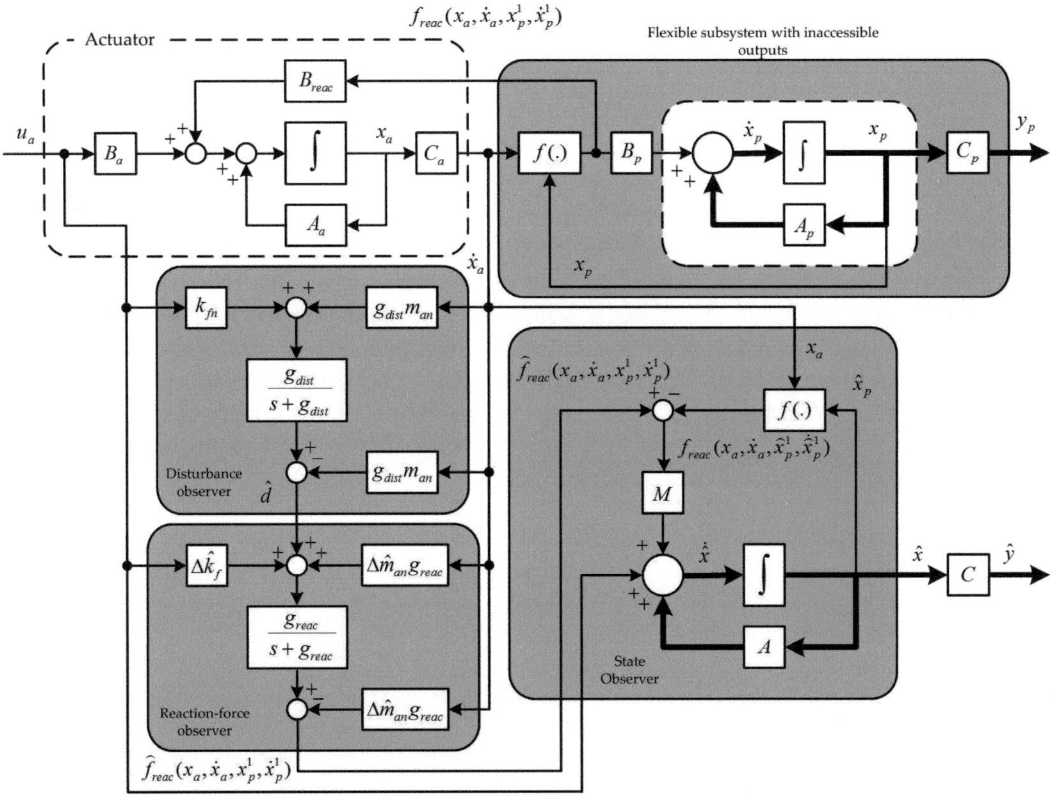

Fig. 8. Action reaction state observer

and vector gains respectively. The design of the action reaction state observer is not trivial since it depends on the function $(f(.))$ which can be determined upon the dynamics of the interacting systems, e.g., if the systems denoted with the subscript (a) and (b) are connected by the mean of a linear elastic element with stiffness k, then the function $f(.)$ can be expressed as $f(.) = k(x_a - x_p)$, here $(x_a - x_p)$ represents the deflection in the elastic element due to the interaction of the two systems. If the two previous systems are connected via an elastic element and energy dissipation element with linear viscous damping c, then the function

$f(.)$ can expressed as $f(.) = k(x_a - x_p) + c(\dot{x}_a - \dot{x}_p)$. if the coupling between the previous systems is defined by a non-linear spring with large deflection (y), the incident reaction force can be defined as $f(.) = k(1 - a^2y^2)y$. Therefore, upon the function $f(.)$, error dynamics of the action reaction state observer can be determined. Obviously, the action reaction state observer gains g_{dist}, g_{reac} and M have to be selected such that the error dynamics of the observer is stable. It can be shown from Fig. 8 that the action reaction state observer consists of three observers in cascade, namely, disturbance force observer, reaction force observer and a Luenberger-like state observer. It is worth noting that the action reaction state observer shown in Fig. 8 is based on first order disturbance observer, first order reaction force observer and $n-2$ order Luenberger-like state observer, where n is the order of the dynamical system (14). The order of the disturbance and reaction force observers depend on the order of the polynomial used to model the disturbance force (11). Therefore, different structures of the action reaction state observer can be designed upon the nature of the disturbance force. Fig. 9 illustrates experimental results of state estimation using the action reaction state observer. This experiment is conducted on a dynamical subsystem (16) with 2 degrees of freedom driven by a single input described by (15). Position and velocity of each degree of freedom of (16) are

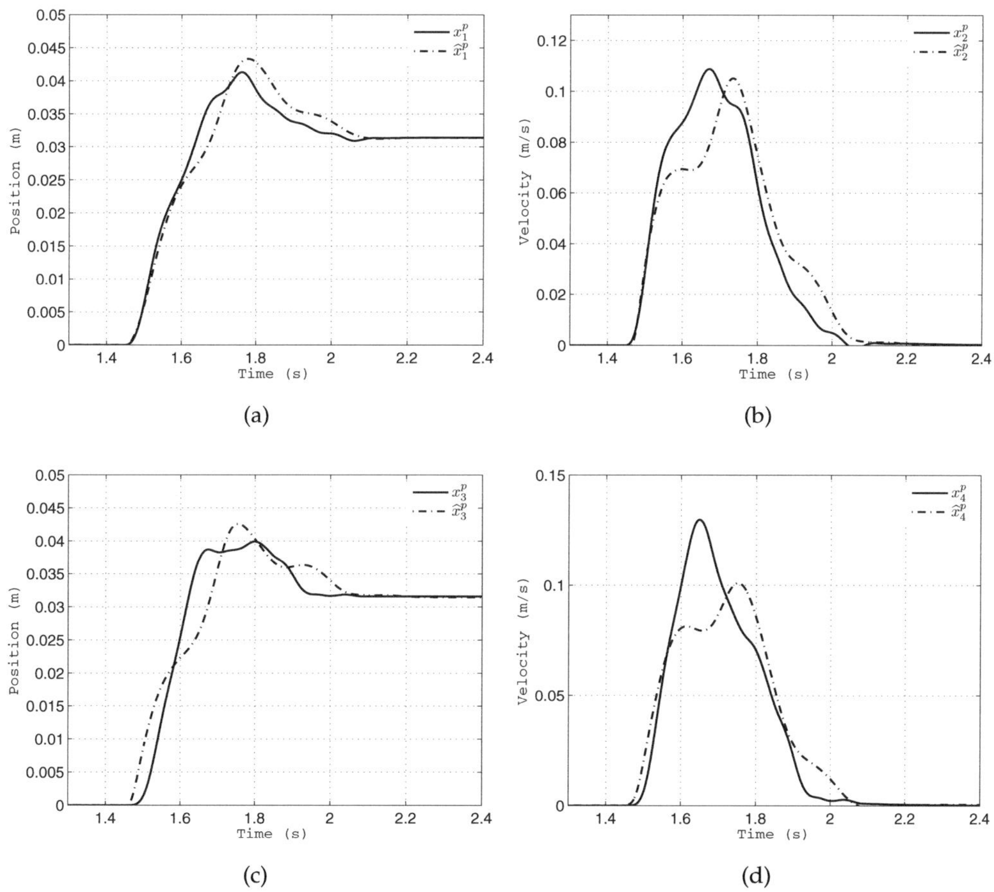

Fig. 9. State estimation experimental result using the action reaction state observer

estimated using the action reaction state observer as shown in Fig. 9 where the estimated states are plotted versus the measured ones in order to verify the validity of the action reaction state observer. The previous experimental result depicted in Fig. 9 indicates that the measurements $(\dot{x}_2)$ and (x_2) can be replaced by their estimates obtained through the action reaction state observer. Therefore, the reaction force observer depicted in Fig. 7 can be further modified by incorporating the action reaction state observer in order to provide estimates of the states $(\dot{x}_2)$ and (x_2). The action reaction state observer is utilized along with the force observer with cut-off frequency (g_{fo}) as shown in Fig. 10 in order not to measure $(\dot{x}_2)$ and (x_2), their estimates $(\dot{\hat{x}}_2)$ and $(\hat{x}_2)$ obtained through the action reaction state observer are used instead. It can be

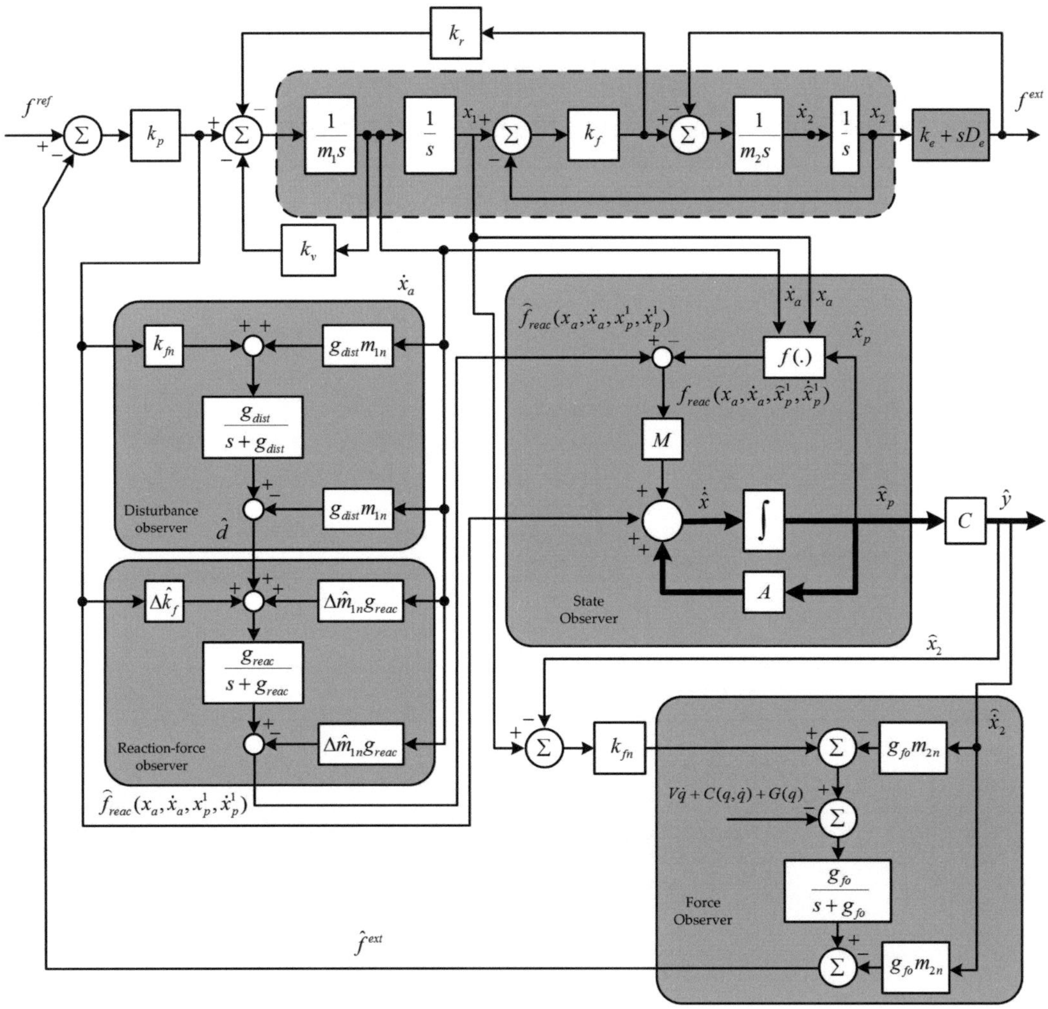

Fig. 10. Action reaction based force control

easily shown from Fig. 10 that the sensorless force control can be realized not only in the absence of the force measurement but also in the absence of the velocity and position of the non-collocated mass in contact with the environment. Therefore, the previous action reaction based force observer avoids the noise associated with force sensing and avoids the noise associated with measuring the end effector (might be non-collocated) position or velocity. The dynamical system depicted in Fig. 10 can be considered as a two mass resonant system in contact with an environment with impedance z_e and the action reaction based force servoing control system shown in the same figure can be considered as a sensorless control system since measurements required to realize the force control are reduced to a single measurement taken from the first degree of freedom, generally this degree of freedom stands for the actuator, i.e., force control can be realized from measurements taken from the actuator whereas the reset of the dynamical system in contact with the environment is kept free from any measurements. Strictly speaking, there exist a measurement taken from the dynamical system in contact with the environment rather than the actuator measurements, namely, the incident reaction forces ($f_{reac}(\mathrm{x_a}, \dot{x}_a, \mathrm{x_p^1}, \dot{x}_p^1)$) which can be estimated as shown previously using a disturbance observer and a reaction force observer in cascade. The central idea of the action reaction based force servoing is based on the well-known action reaction principle of dynamics. The dynamical system we considered in the previous analysis is a two mass resonant system, the second degree of freedom is in contact with an environment. Therefore, whenever the second degree of freedoms (end-effector) comes into contact with the environment the instantaneous reaction can be observed if the velocity of the second degree of freedom is measured as shown in Fig. 7. In order to keep the second degree of freedom or the end effector even free from this velocity or position measurement, the action reaction state observer is utilized to provide the force observer with the velocity and/or the position estimates of the second degree of freedom, thus realizing the force control from measurement taken from the active degree of freedom of the system, namely the actuator.

Application of the sensorless force control techniques and the action reaction based force servoing are oriented toward systems at which measurements cannot be made such as microsystems and micromanipulation operations. In these specific application, sensor utilization is commonly believed to be one of the major reasons for not having these systems automated till now as it was reported in (Savia & Koivo, 2009). A micromanipulation system is shown in Fig. 11 which consists of a master side and a remote slave side used for biological cell characterization. The end effector of the slave robot is used in manipulating, injecting and characterization of biological cells through scaled bilateral teleoperations. Such applications requires measuring the interaction forces in order to achieve safe interaction with these fragile cells or any similar objects. However, the force sensing is very problematic due to the reasons listed and discussed in the previous sections. In addition, due to the very limited workspace, having force sensor embedded to the end effector of the remote slave robot is hard to be achieved. The current ongoing research in the area of microsystems and micromanipulation is based on utilization of the AFM probes which are expensive and have the same typical problem listed in the previous sections. Many authors on the other hand tried to utilize visual feedback in order to realize the interaction forces between the end-effector and biological cells, however this process cannot provide online estimated force feedback which is needed to realize the force control (Huang et al., 2007). Therefore, the sensorless force control techniques proposed in this section can be utilized in such application in order to spare the remote slave robot from having force sensors attached to it. In addition, utilization of these sensorless force control techniques reduces the complicity of the entire control system.

In the realization of the force control, the estimated forces are used instead of the measured forces. However, a verification is always required in order to demonstrate that the estimated forces are converging to the actual ones. This can be done by comparing the actual forces

Fig. 11. Micro-system experimental setup for sensorless force control operations on biological cells

with the estimated ones. Similar to the experimental result conducted to verify the validity of the action reaction state observer, another experiment has to be performed in order to compare the estimated forces obtained through the action reaction force observer depicted in Fig. 10. Experimentally, a lumped mass spring system with three degrees of freedom is used in order to verify the validity of the action reaction based force observer. The lumped mass spring system has a single input from which measurements can be taken, namely the velocity of the actuator along with the current reference input. The single input is labeled with the symbol (A) as depicted in Fig. 12, at the end effector of the lumped system another actuator (B) is mounted and conceptually considered as an environment with impedance (z_e). Thus using actuator or the environment (B) the impedance can be varied and experiments can be conducted easily on a soft and hard environments by changing the impedance of actuator (B). The sensorless force control can be achieved by forcing the end effector of the lumped mass spring system (The last non-collocated mass) to exert the desired force on the environment (B) without using force sensors, the force servoing sensorless control framework depicted in Fig. 7 requires measuring the velocity of the end effector and using it as basis of the force observer, whereas, the action reaction based force servoing approach depicted in Fig. 10 does not require taking this measurement or any measurement from the dynamical subsystem attached to the single input (A). This means that the sensorless force control is realized from measurements taken from actuator (A), whereas the dynamical subsystem is kept free from any measurement and attached sensors. The parameters of this experiment are included in Table. 1. Optical encoders with 1μm resolution are attached to each degree of freedom as depicted in Fig. 12 in order to compare the estimated states with the actual ones. These encoder signal are just used for the sake of verification only.

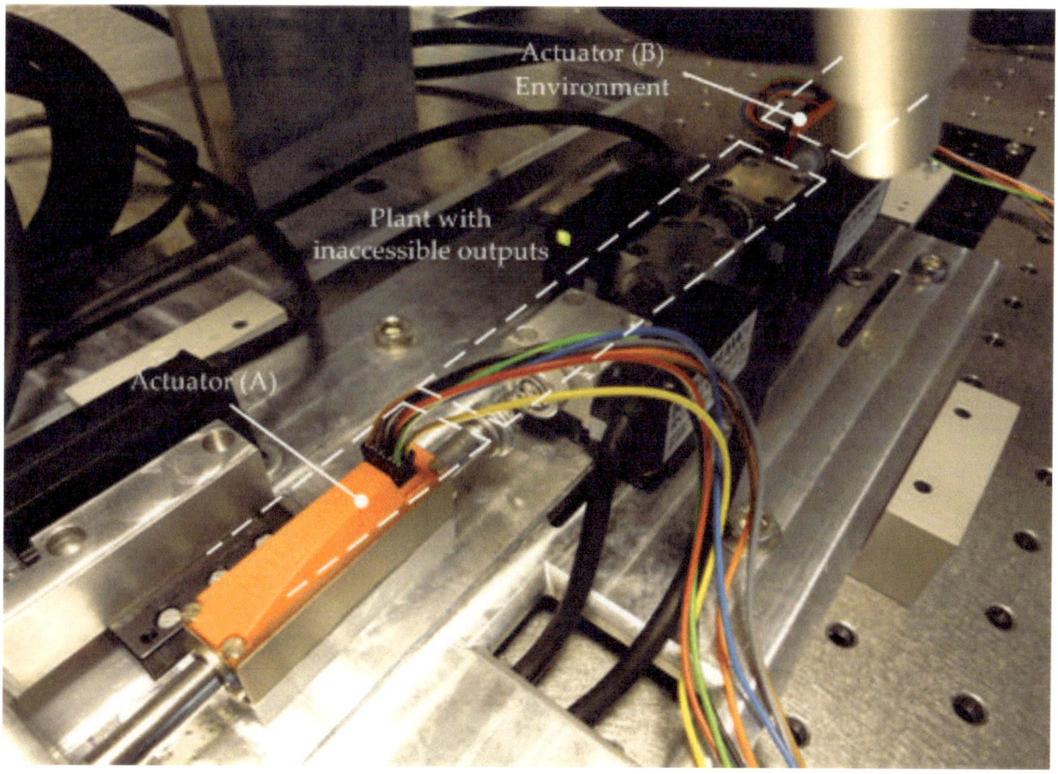

Fig. 12. Sensorless force control experimental setup

The experimental results of the action reaction external force estimation are depicted in Fig. 13 and Fig. 14 at which the estimated interaction force using the action reaction based force observer is compared with the measured interaction forces. The experiments were performed in different frequencies and amplitudes and in each case the actual interaction force was compared with the estimated one in order to verify the validity of the proposed action reaction based force observer. The experimental results shown in Fig. 13 and Fig. 14 indicate that the estimated interaction force ($\widehat{f^{ext}}$) can further be used in the realization of the sensorless force control as depicted in Fig. 10. Fig. 14 illustrates the force estimation experimental result at higher frequency of the external interaction force, this result shows that the observer satisfactory operate in a wide frequency range of the dynamical system.

The force observer depicted in Fig. 10 consists of four observers in cascade, namely the disturbance force, reaction force, the Luenberger-like state observer and the force observers. Therefore, the action reaction based force observer has four degrees of freedom or scaler and vector gains which have to be selected properly to accomplish stable convergence of the

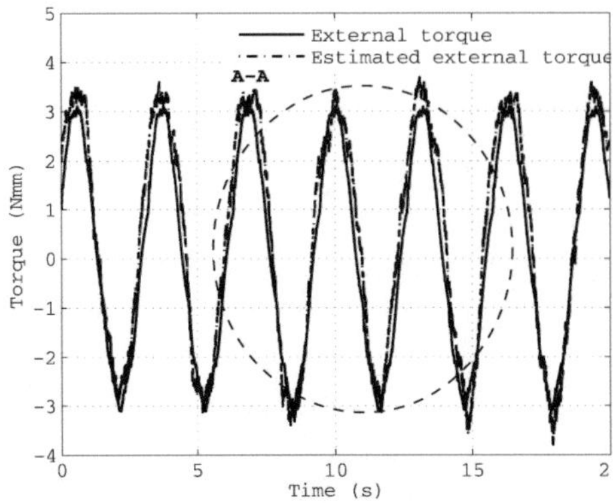

(a)

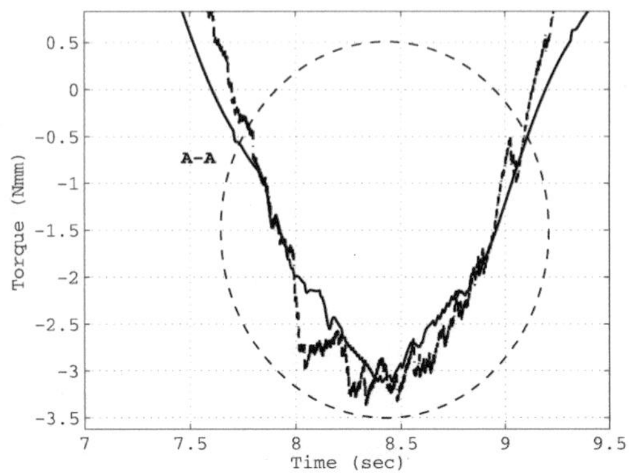

(b)

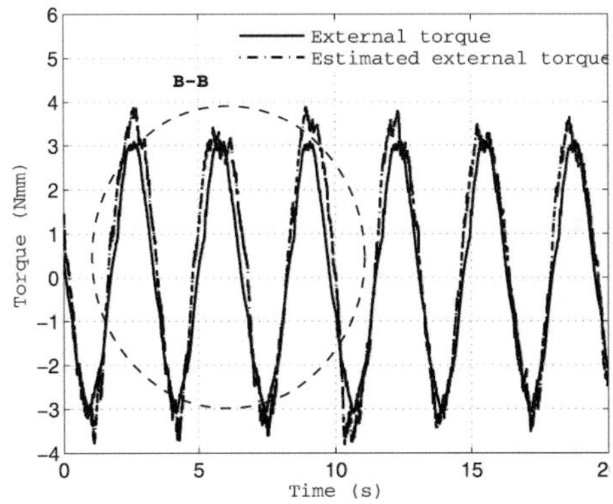

(c)

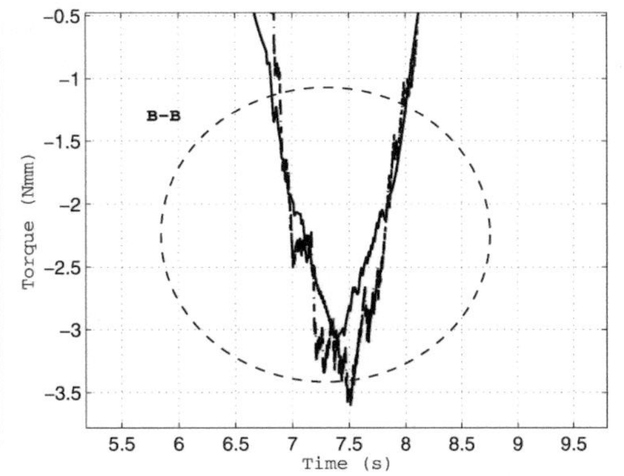

(d)

Fig. 13. Action reaction based force estimation experimental results

Actuator force constant	k_{fn}	6.43	N/A
Actuator Nominal mass	m_{an}	0.059	kg
Lumped masses	$m_{1,2,3}$	0.019	kg
Identified spring constants	$k_{1,2,3}$	503.96	N/m
Identified viscous damping coefficients	$c_{1,2,3}$	0.262	Ns/m
Reaction force observer gain	g_{reac}	628	rad/s
Disturbance observer gain	g_{dist}	628	rad/s
Force observer gain	g_{fo}	628	rad/s
Low-pass filter gain	g_{l}	1000	rad/s
Sampling time	T_s	1	ms

Table 1. Experimental and simulation parameters

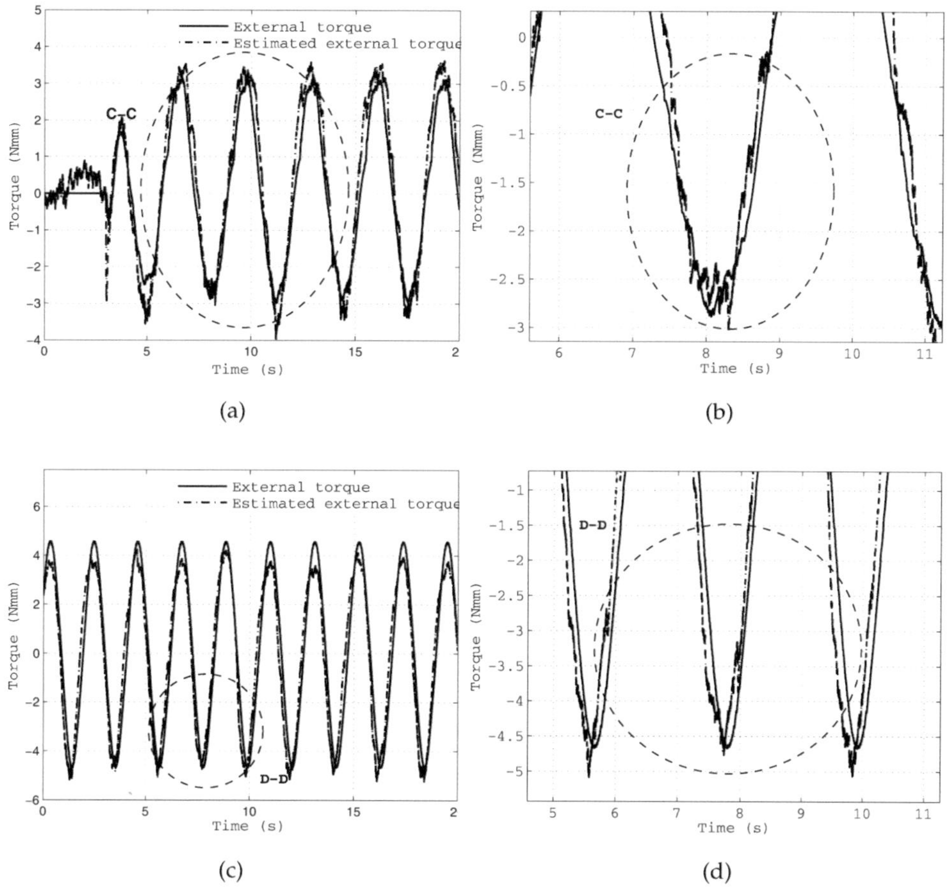

Fig. 14. Action reaction based force estimation experimental results

estimated interaction force to the actual one. The experimental values of the scaler observers gains (g_{dist}, g_{reac} and g_{fo}) are included in Table. 1 while the Luenberger-like state observer gain vector (M) can be determined using the following characteristic equation

$$|sI - (I - cML)^{-1}(A + kML)| = (s - \mu_1)\dots(s - \mu_n) \tag{20}$$

$$L = \begin{bmatrix} 0 \; 0 \; 1 \; 0 \dots 0 \end{bmatrix}$$

where $|sI - (I - cML)^{-1}(A + kML)|$ is the desired characteristic polynomial and $\mu_1, \dots, \mu_n$ are the desired observer poles. The previous equation can be determined by subtracting the state space equation of the dynamical system from the action reaction state observer equation and further performing few algebraic manipulation to obtain a dynamical equation for the estimation error ($e = x - \hat{x}$). The error dynamic therefore can be shown as

$$\dot{e} = (I - cML)^{-1}(A + kML)e = \mathcal{A}e \tag{21}$$

Here, the function $f(.)$ is assumed to be a combination of a linear spring with stiffness k and linear energy dissipating element with viscous damping (c), A $\in \mathbb{R}^{n \times n}$ is the system matrix, I $\in \mathbb{R}^{n \times n}$ is the identity matrix and L $\in$ R$^{1 \times n}$. The previous estimation error dynamical equation is not trivial, it is rather depending on the interaction force model between the dynamical subsystem and the input of the system, i.e., in driving the error dynamical equation, the interaction force model have to be determined and further used in the derivation of the estimation error dynamics.
The Luenberger-like state observer vector gain (M) can be determined by directly substituting the desired observer poles in (20). On the other hand, the scaler observer gains have to be selected in a way that satisfies two criterions, i.e., convergence time and noise attenuation since disturbance force, reaction force and the environmental interaction force a observer through low-pass filters. Therefore, the tradeoff between convergence time and noise attenuation restrict the selection of these scaler observers gains. It is worth noting that the action reaction force observer or state observer increases the flexibility and the range of these scaler gains since the noise associated with sensors is minimized as the number of measurements is reduced. The action reaction state observer or the action reaction force observers depends on a single measurement from the actuator along with the reference current input. Therefore, a single sensor has to be used in the realization of these state or force observers. The reference current signal is supplied through the computer control. Therefore, it is not a measurement it is rather a reference supplied signal. The pervious argument indicates that the action reaction force observer is depending on a single measurement from the dynamical system, the sensor noise is therefore reduced as the number of sensors utilized in the entire control system is minimized. It is worth noting that throughout the previous experiments the velocity measurement was obtained by differentiating the position encoder signal through a low-pass filter with cut-off frequency g_l, this explains the noise nature of the estimated interaction force depicted in Fig. 13 and Fig. 14. Therefore, it is recommended to obtain this velocity signal through an optimal filter such as Kalman filter rather than using a simple low-pass filter to attenuate the noise effect. Nevertheless, the results obtained in this chapter demonstrate the validity of the proposed sensorless force estimation techniques which can be further enhanced by utilizing optimal filter to reduce the sensor noise associated with the velocity measurement required to realize the action reaction state and force observers.

5. Conclusion

Sensorless force/torque control can be realized through utilization of the force observer techniques along with the action reaction state observer in order to provide an action reaction based force observer which can be used in the realization of sensorless force control. The action reaction based force observer does not require having any measurement from the dynamical system in contact with the environment except a single measurement from the input (actuator). The action reaction based force observer consists of four observers in cascade. The scalers and vector gains of these observers have to be selected such that convergence of the estimated interaction force to the actual one is guaranteed. The central idea of the action reaction state observer or the action reaction force observer is based on the realization of a natural feedback or a feedback-like force from the dynamical system which interact with the environment. This feedback-like force is then used in the design of a state and force observer by injecting the estimated reaction force onto the Luenberger-like state observer structure, thus allowing the estimated states or forces to converge to the actual ones through a stable error dynamics that can be accomplished by the proper selection of the observer scaler and vector gains. The action reaction based force servoing is introduced based on the action reaction state observer and the force observer. This force control technique is considered as a sensorless force control technique since force sensors are not utilized in the realization of the force control. In addition, velocity measurement of the non-collocated end effector in contact with the environment does not have to be measured. Experiments are conducted on a single input flexible lumped system with multi degree of freedom in contact with an environment with varying impedance. The estimation results of the states and the interaction forces were satisfactory showing fast convergence over a wide bandwidth of the dynamical system. Since the action reaction force observer depends on a single measurement from the system along with the reference current input, the associated noise problem is minimized and this allows the observer to work in a wide bandwidth. This in turn allows increasing the cut-off frequencies associated with each observer low-pass filter in order to reduce the convergence time of each observer of the action reaction state or force observer.

6. References

Huang, H., Sun, D., Mills, J. K. & Li, W. J. (2007). Visual-based impedance force control of three-dimensional cell injection system, *Proceedings of IEEEInt. Conf. on Robotics and Automation, ICRA'07*, IEEE, Roma, pp. 4196–4201.

Iida, W. & Ohnishi, K. (2003). Sensorless force control with force error observer, *Proceedings of IEEEInt. Conf. on Industrial Technology, ICIT'11*, IEEE, Maribor, pp. 157–162.

Karunakar, S. B. & Goldenberg, A. A. (1988). Sensorless action-reaction-based residual vibration suppression for multi-degree-of-freedom flexible systems, *Proceedings of IEEEInt. Symp. on Intelligent Control, ISIC'98*, IEEE, Arlington, VA, pp. 412 – 417.

Katsura, S., Irie, K. & Ohishi, K. (2008). Wideband force control by position-acceleration integrated disturbance observer, *IEEE Trans. Ind Electron* 55.(4.): 1699–1706.

Katsura, S., Matsumoto, Y. & Ohnishi, K. (2007). Modeling of force sensing and validation of disturbance observer for force control, *IEEE Trans. Ind Electron* 54.(1.): 530–538.

Katsura, S. & Ohnishi, K. (2007). Force servoing by flexible manipulator based on resonance ratio control, *IEEE Trans. Ind Electron* 54.(1.): 56–67.

Khalil, I. S. M. & Sabanovic, A. (2010a). Action-reaction based parameters identification and states estimation of flexible systems, *Proceedings of IEEEInt. Conf. Industrial Technology, ISIE'10*, IEEE, PARI, pp. 46–51.

Khalil, I. S. M. & Sabanovic, A. (2010b). Sensorless action-reaction-based residual vibration suppression for multi-degree-of-freedom flexible systems, *Proceedings of IEEEInt. Conf. Industrial Technology, IECON'10*, IEEE, Glendale, AZ, pp. 1633–1638.

Khalil, I. S. M. & Sabanovic, A. (2011). A novel state observer for dynamical systems with inaccessible outputs, *Proceedings of IEEEInt. Conf. Mechatronics, ICM'11*, IEEE, Istanbul, pp. –.

Khalil, I. S. M. & Sabanovic, A. (n.d.). Action reaction state observer for single input systems with compliance, *IEEE Trans. Ind Electron* .

Li, Y. F. & Chen, X. B. (1998). On the dynamic behavior of a force/torque sensor for robots, *IEEE Trans. Ind Electron* 47.(1.): 304–308.

Murakami, T., Yu, F. & Ohnishi, K. (1993). Torque sensorless control in multidegree-of-freedom manipulator, *IEEE Trans. Ind Electron* 40.(2.): 259–265.

O'Connor, W. J. (1998). Position control of flexible robot arms using mechanical waves, *ASME J. Dyn. Syst., Meas. Control* 120.(3.): 334–339.

O'Connor, W. J. (2003). A gantry crane problem solved, *ASME J. Dyn. Syst., Meas. Control* 120.(4.): 569–576.

O'Connor, W. J. (2007a). Theory of wave analysis of lumped flexible systems, *Proceedings of IEEEInt. Conf. American Control, ACC'07-NEW YORK*, IEEE, NEW YORK, pp. 4215–4220.

O'Connor, W. J. (2007b). Wave-based analysis and control of lump-modeled flexible robots, *IEEE Trans. Robotics* 23.(.): 1552–3098.

Ohnishi, K., Shibata, M. & Murakami, T. (1996). Motion control for advanced mechatronics, *IEEE/ASME Trans. Mechatronics* 1.(1.): 56–67.

Savia, M. & Koivo, H. N. (2009). Contact micromanipulation-survey of strategies, *IEEE/ASME Trans. Mechatronics* 14.(4.): 504–514.

Part 5

Direct Torque Control

5

Speed Estimation Using Extended Filter Kalman for the Direct Torque Controlled Permanent Magnet Synchronous Motor (PMSM)

M. S. Merzoug and H. Benalla
LEC- Research Laboratory de Constantine
Department of Electrical Engineering
Constantine,
Algeria

1. Introduction

Permanent magnet (PM) synchronous motors have attracted increasing interest in recent years for industrial drive application. The high efficiency, high steady state torque density and simple controller of the PM motor drives compared with the induction motor drives make them a good alternative in certain applications. Moreover, the availability of low-cost power electronic devices and the improvement of PM characteristics enable the use of PM motors even in some more demanding applications. Advantages of PMSM include low inertia, high efficiency, high power density and reliability. Because of these advantages, PMSM are indeed excellent for use in high-performance servo drives where a fast and accurate torque response is required. In PMSM drives, the electromagnetic torque is usually controlled indirectly via the stator current components in a reference frame fixed to the rotor flux field. The idea of combining the advantages of DTC and PMSM into a highly dynamic drive appeared in the literature in the late 1990's. In the past decade several authors have proposed ways to adapt DTC to work with PMSM.
Moreover DTC uses no current controller and no motor parameters other than the stator resistance, which yields a faster torque response and lower parameter dependence than with field oriented control.
Recently, motor drive systems without electromechanical sensors, so called 'sensorless drives', have gained increasing popularity in industrial applications because of inherent drawbacks of electromechanical sensors. In general, electromechanical sensors are used to obtain speed or position information of motors. A drawback of these sensors is performance degradation due to vibration or humidity .
Consequently, the extended Kalman filter is considered to be the best solution for the speed estimation of an induction motor. The extended Kalman filter has been applied to the direct torque control control system .

2. The Model of PMSM

The electrical and mechanical equations of the PMSM in the rotor reference (d-q) frame are as follows [6]:

$$\begin{cases} \frac{d}{dt}I_d = -\frac{R_s}{L_d}I_d + \omega_r \frac{L_q}{L_d}I_q + V_d \\ \frac{d}{dt}I_q = -\frac{R_s}{L_q}I_q - \omega_r \frac{L_d}{L_q}I_d - \frac{1}{L_q}\varphi_f \omega_r + V_q \end{cases} \tag{1}$$

$$\varphi_d = L_d I_d + \varphi_f \tag{2}$$

$$\varphi_q = L_q I_q \tag{3}$$

And the electromagnetic torque T_e is given by:

$$T_e = \frac{3}{2}P[(L_d - L_q)I_d I_q + I_q \varphi_f)] \tag{4}$$

The equation for the motor dynamics, on the other hand is

$$T_e - T_L - T_f = J\frac{d\Omega}{dt} \tag{5}$$

Figure. 1, presents the model of PMSM in d-q axis

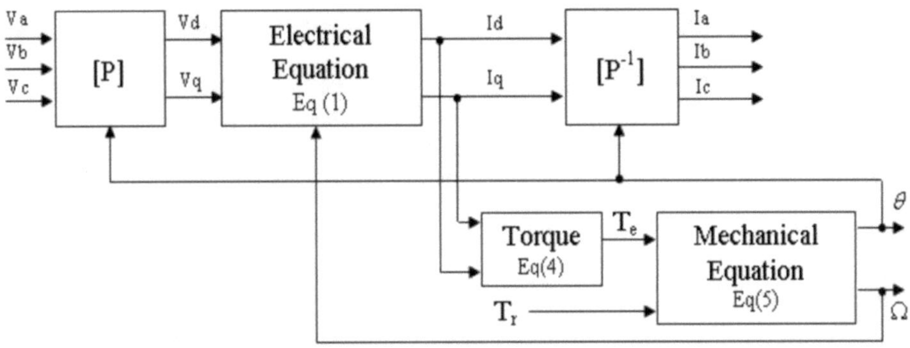

Fig. 1. Model of PMSM in d-q axis.

3. Direct torque control system

Since M. Depenbrock and I. Takahashi proposed Direct Torque Control (DTC) for induction machines in the middle of 1980's, more than one decade has passed. It is getting more and more popular nowadays. The basic idea of DTC for induction motor is to control the torque and flux linkage by selecting the voltage space vectors properly, which is based on the relationship between the slip frequency and torque.

3.1 Estimation of the torque and the flux linkage

The basic principle of the DTC is to select proper voltage vectors using a pre-defined switching table. The selection is based on the hysteresis control of the stator flux linkage and the torque. In the basic form the stator flux linkage is estimated with:

$$\varphi_s(t) = \int_0^t (Vs - R_s I_s)dt + \varphi_{s0} \tag{6}$$

Let us replace the estimate of the stator voltage with the true value and write it as:

$$V_S(S_A;S_B;S_C) = \frac{2}{3}U_0\left(S_A + S_B e^{j2\pi/3} + S_C e^{j4\pi/3}\right) \tag{7}$$

Sa, Sb and Sc represent the states of the three phase legs 0 meaning that the phase is connected to the negative and 1 meaning that the phase is connected to the positive leg.
The stator current space vector is calculated from measured currents i_a, i_b, i_c :

$$i_S = \frac{2}{3}\left(i_a + i_b e^{J2\pi/3} + i_c e^{j4\pi/3}\right) \tag{8}$$

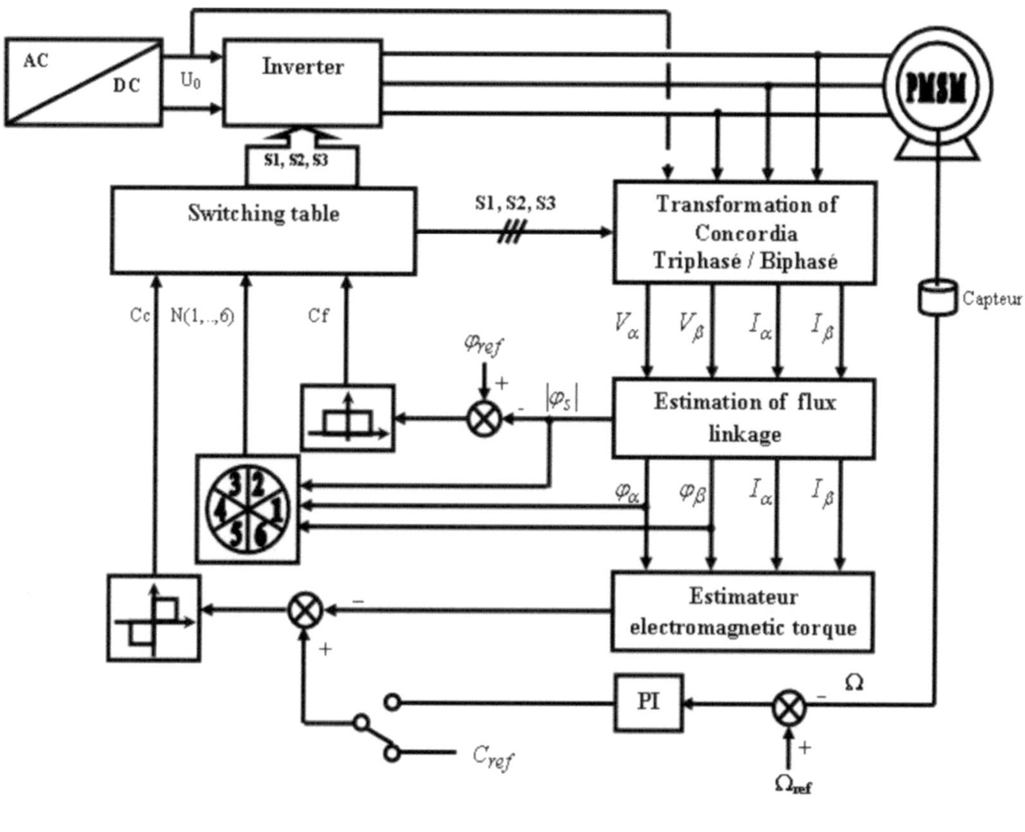

Fig. 2. System diagram of a typical DTC PMSM drive system.

The voltage vectors obtained this way are shown in Figure 3.

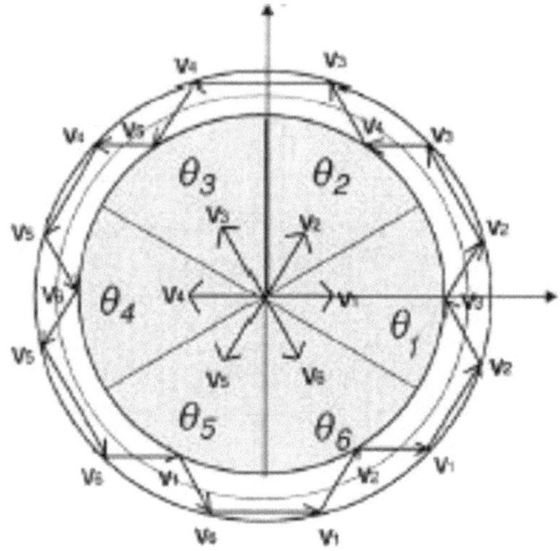

Fig. 3. Voltage vectors for DTC.

The composite α and β of vector φ_s can be obtained:

$$\begin{cases} \varphi_{S\alpha} = \int_0^t \left(V_{S\alpha} - R_S I_{S\alpha}\right) dt \\ \varphi_{S\beta} = \int_0^t \left(V_{S\beta} - R_S I_{S\beta}\right) dt \end{cases} \tag{9}$$

Stator flux linkage written:

$$\varphi_s = \sqrt{\left(\varphi_{S\alpha}\right)^2 + \left(\varphi_{S\beta}\right)^2} \tag{10}$$

The angle θ_S is equal to:

$$\theta_S = arctg \frac{\varphi_{S\beta}}{\varphi_{S\alpha}} \tag{11}$$

The torque can then be estimated with:

$$T_e = \frac{3}{2} P(\varphi_{S\alpha} I_{S\beta} - \varphi_{S\beta} I_{S\alpha}) \tag{12}$$

The voltage vector plane is divided into six sectors so that each voltage vector divides each region into two equal parts. In each sector, four of the six non-zero voltage vectors may be used. Also zero sectors are allowed. All the possibilities can be tabulated into a switching table (Table I). The output of the torque hysteresis comparator is denoted as τ , the output of the flux hysteresis comparator as ϕ and the flux linkage sector is denoted as θ .

θ, τ, ϕ		θ_1	θ_2	θ_3	θ_4	θ_5	θ_6
$\phi = 1$	$\tau = 1$	V_2	V_3	V_4	V_5	V_6	V_1
	$\tau = 0$	V_7	V_0	V_7	V_0	V_7	V_0
	$\tau = -1$	V_6	V_1	V_2	V_3	V_4	V_5
$\phi = 0$	$\tau = 1$	V_3	V_4	V_5	V_6	V_1	V_2
	$\tau = 0$	V_0	V_7	V_0	V_7	V_0	V_7
	$\tau = -1$	V_5	V_6	V_1	V_2	V_3	V_4

Table 1. Switching Table Presented By Takahashi and Noguchi.

3. Kalman filter

The Kalman filter was developed by R.E. Kalman in 1960. Due to advances in the development of digital computing, the Kalman filter is a subject of extensive research and application. Kalman filtering has been applied in the areas of aerospace, navigation, manufacturing, and many others. [9]

The Kalman filter provides a means for inferring missing information from indirect (and noisy) measurements. It provides the optimal (minimum variance) state estimate when the dynamic system is linear and the statistical characteristics of the various noise elements are know

The EKF is an optimal recursive estimation algorithm based on the least-square sense for estimating the states of dynamic nonlinear systems. That is, it is an optimal estimator for computing the conditional mean and covariance of the probability distribution of the state of a nonlinear stochastic system with uncorrelated Gaussian process and measurement noise.

Since the state models are nonlinear, the EKF can be applied to estimate state variables. In this case, the back-EMF is considered as a state variable. Nonlinear discrete models with white noise are given as follows:

$$\begin{cases} x(k+1) = F(x(k), u(k)) + w(k) \\ y(k) = h(x(k)) + v(k) \end{cases} \tag{13}$$

w and v are the system and measurement noise.

For linearization process in the model, the partial derivative is introduced and discrete state models are: [10]

$$F(k) = \left. \frac{\partial f(x(k), u(k))}{\partial x^T(k)} \right|_{x(k)=\hat{x}(k/k)} \tag{14}$$

$$H(k) = \left. \frac{\partial f(x(k))}{\partial x(k)} \right|_{x(k)=\hat{x}(k/k)} \tag{15}$$

Estimation of an error covariance matrix

$$P^{-}(k+1) = F(k).P(k).F^T + Q \tag{16}$$

Computation of a Kalman filter gain

$$K(k+1) = P^{-}(k+1)H^{T}(k)\left[H(k)P^{-}(k+1)H(k)^{T} + R\right]^{-1} \quad (17)$$

Update of a error covariance matrix

$$P(k+1) = \left(I - K(k+1).H(k)\right)P^{-}(k+1) \quad (18)$$

State estimation

$$\hat{X}(k+1) = \hat{X}(k) + K(k+1)\left[y(k+1) - h\left(\hat{X}(k+1)\right)\right] \quad (19)$$

Where:

$P^{-}(k+1)$ is a priori error covariance matrix

Q and R respectively and are independent from the system state [4].

4. EKF implementation for a PMSM

The extended Kalman filter implementation for a PMSM motor requires three basic steps.

- Continuous PMSM model
- Discretization of the PMSM motor model
- Simulation

4.1 Continuous PMSM model

The model of PMSM in the d-q reference frame can be written:

$$\begin{cases} L_d \dfrac{d}{dt} I_d = V_d - R_s I_d + P\Omega L_q I_q \\ L_q \dfrac{d}{dt} I_q = V_q - R_s I_q - P\Omega(L_d I_d + \varphi_f) \\ J\dfrac{d\Omega}{dt} = T_e - T_L - f\Omega \\ \dfrac{d\theta}{dt} = \Omega \end{cases} \quad (20)$$

The system can be modeled as :

$$\frac{d}{dt}\begin{bmatrix} I_d \\ I_q \\ \Omega \\ \theta \end{bmatrix} = \begin{bmatrix} -\dfrac{R_s}{Ld} & P\Omega\dfrac{L_q}{Ld} & 0 & 0 \\ -P\Omega\dfrac{L_d}{L_q} & -\dfrac{R_s}{L_q} & -P\dfrac{\varphi_f}{L_q} & 0 \\ 0 & 0 & -\dfrac{f}{J} & 0 \\ 0 & 0 & 0 & 1 \end{bmatrix}\begin{bmatrix} I_d \\ I_q \\ \Omega \\ \theta \end{bmatrix} + \begin{bmatrix} \dfrac{1}{L_d} \\ \dfrac{1}{L_q} \\ -\dfrac{1}{J} \\ 0 \end{bmatrix}\begin{bmatrix} V_d \\ V_q \\ T_L \\ 0 \end{bmatrix} \quad (21)$$

We are measuring current I_d and I_q.

$$\begin{bmatrix} I_d \\ I_q \end{bmatrix} = \begin{bmatrix} 1 & 0 & 0 & 0 \\ 0 & 1 & 0 & 0 \end{bmatrix} \begin{bmatrix} I_d \\ I_q \\ \Omega \\ \theta \end{bmatrix} \tag{22}$$

4.2 Discretization of the PMSM model

The corresponding discrete time model is given by:

$$X_{(k+1)} = A_d . X_{(k)} + B_d . u_{(k)} \tag{23}$$

$$y_{(k+1)} = C_d . X_{(k)} \tag{24}$$

The conversion is done by the following approximation;

$$\begin{cases} A_d = e^{At} = I + ATs \\ B_d = \int_0^t e^{A\xi} B d\xi = B.Ts \\ C_d = C \end{cases} \tag{25}$$

We are assuming that the sampling -time T_S is very small compared to the system dynamics. The discrete model of the PMSM motor is given as:

$$f(x(k),u(k)) = \begin{bmatrix} (1 - T_s \frac{R_s}{L_d}) I_d + p\Omega \ T_s \frac{L_q}{L_d} I_q + T_s \frac{1}{L_d} v_d \\ (-p\Omega T_s \frac{L_d}{L_q}) I_d + (1 - T_s \frac{R_s}{L_q}) I_q - T_s \frac{\varphi_f}{L_q} p\Omega + T_s \frac{1}{L_q} v_q \\ PT_s \frac{L_d - L_q}{J} I_q I_d + pT_s \frac{\varphi_f}{J} I_q + (1 - T_s \frac{f}{J}) \Omega - T_s \frac{1}{J} T_L \\ \theta + T_s \Omega \end{bmatrix} \tag{26}$$

4.3 Simulation and implementation

The matrices linearization F and H, we can linearize the system in every moment of operation. They are given as follows

$$F(k) = \begin{bmatrix} 1 - Ts\frac{R_s}{Ld} & P\Omega Ts\frac{L_q}{Ld} & Ts\frac{PL_q}{Ld} I_q & 0 \\ -P\Omega Ts\frac{L_d}{L_q} & 1 - Ts\frac{R_s}{L_q} & -\frac{Ts}{L_q}\left(PL_d I_d + P\varphi_f\right) & 0 \\ PTs\frac{Ld - Lq}{J} I_q & PTs\left(\frac{Ld - Lq}{J} Id + \frac{\varphi}{J}\right) & 1 - Ts\frac{f}{J} & 0 \\ 0 & 0 & Ts & 1 \end{bmatrix} \tag{27}$$

$$h = \begin{bmatrix} I_d \\ I_q \end{bmatrix} \tag{28}$$

$$\frac{\partial H}{\partial X} = \begin{bmatrix} 1 & 0 & 0 & 0 \\ 0 & 1 & 0 & 0 \end{bmatrix} \tag{30}$$

4.4 Observer torque load

The method proposed by Pioufle can be estimated the torque load, Figure 4, illustrates the principle of the estimate. [11]

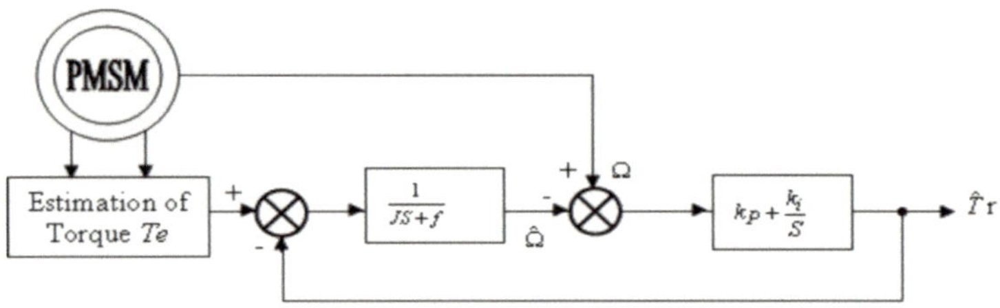

Fig. 4. Estimation load torque.

The error between the speed measured and the estimated speed entry is presented as a regulator PI whose output is

$$\hat{T}_L = \frac{1 + \frac{kp}{ki} S}{1 + \frac{1 + kp}{ki} S + \frac{1}{ki} S^2} T_L \tag{31}$$

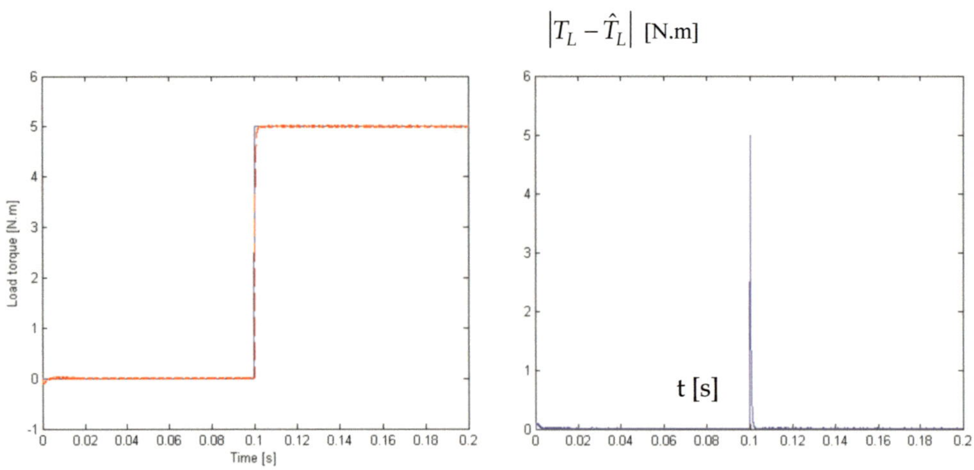

Fig. 5. Estimation load torque.

5. Sensitivity study and simulation results

Computer simulations have been carried out in order to validate the effectiveness of the proposed scheme. The speed current and rotor position responses are observed under various operating conditions such as change in reference speed, step change in load and parameter variation.

6. Simulation results

Extensive simulations have been performed using Matlab/ Simulink Software to examine control algorithm of the DTC applied for PMSM presented in Figure 6.

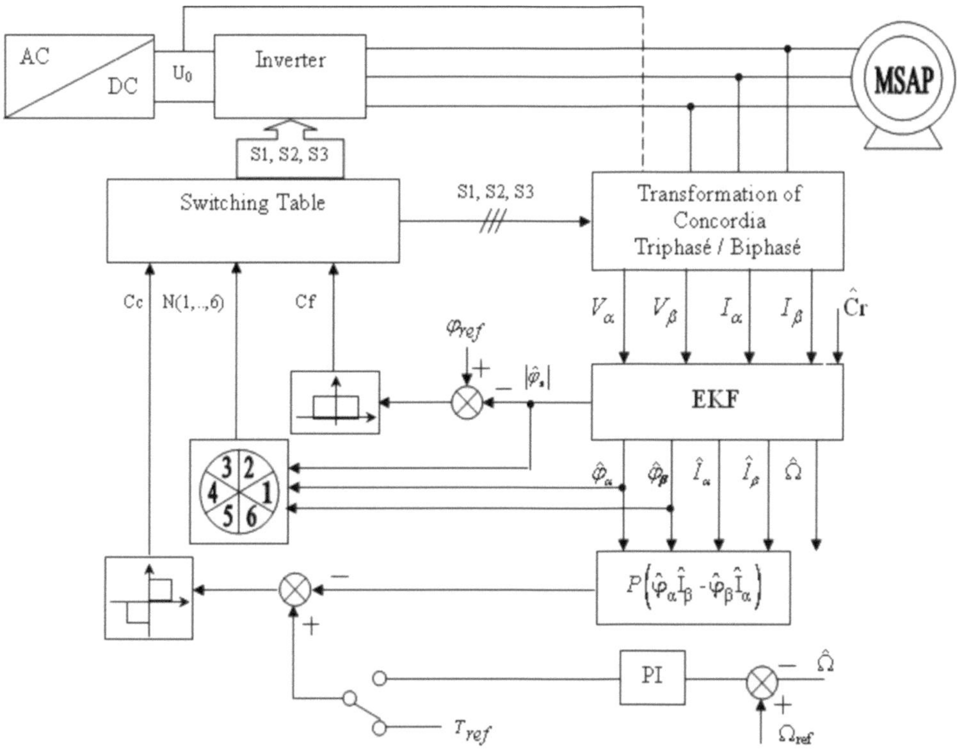

Fig. 6. Basic DTC for sensorless PMSM drive.

6.1 Regulation speed

The simulation results of the DTC where to apply a torque load equal to 5 N.m at t = 0.1s and a reference speed equal to 100 rad /s.

Figure 7, shows the actual and estimated speed of the proposed PMSM sensorless scheme, these responses illustrate high performance of the proposed EKF observer during transients and steady state.

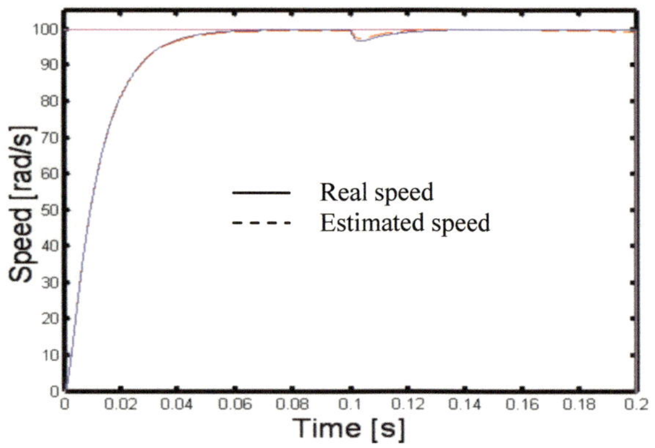

Fig. 7. Actual and estimated speed.

Figure 8, present the speed zoom with and without torque compensation.

Figure 9, presents the estimated electromagnetic torque. The torque follows perfectly the value of record with negligible effect on the speed that is rapidly restored to its reference.

Figure 10, presents the estimated flux, we also note that the flow is not affected by the variation of the load. The flux reaches its reference control with no overshoot.

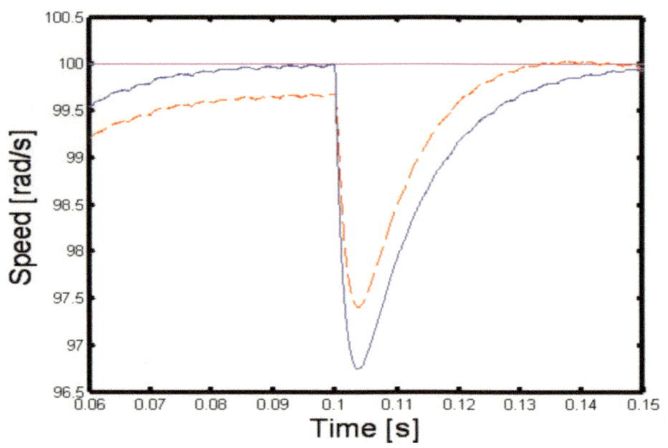

Fig. 8. Speed zoom.

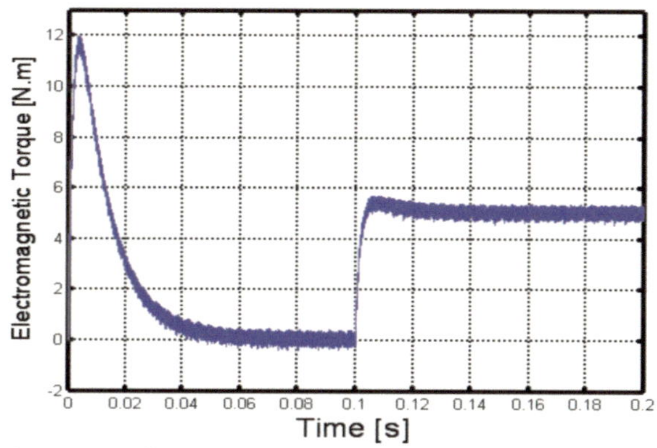

Fig. 9. Estimated electromagnetic torque.

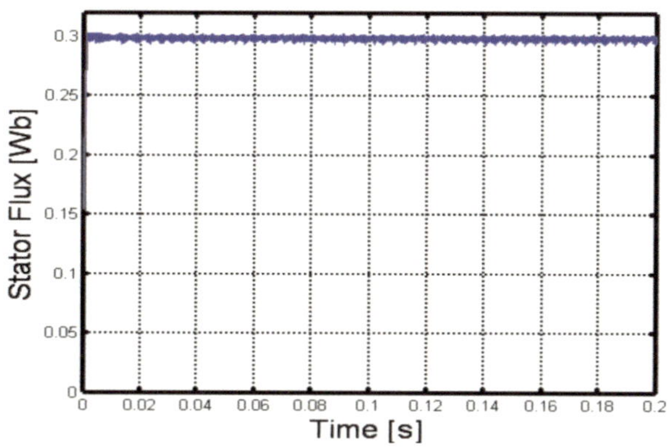

Fig. 10. Estimated stator flux.

6.2 Inversion of the speed

To test the robustness of technical command at the reverse direction of rotation, it introduced a change in record speed reference +100 rad /s to -100 rad /s time t=0.1s after a torque load equal to 3 Nm, Figure 11.

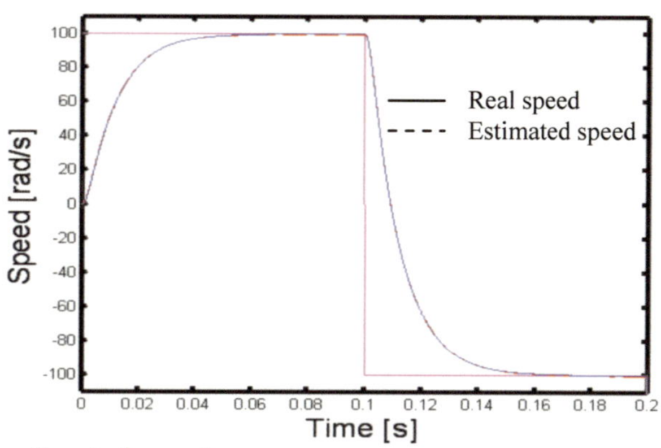

Fig. 11. Actual and estimated speed.

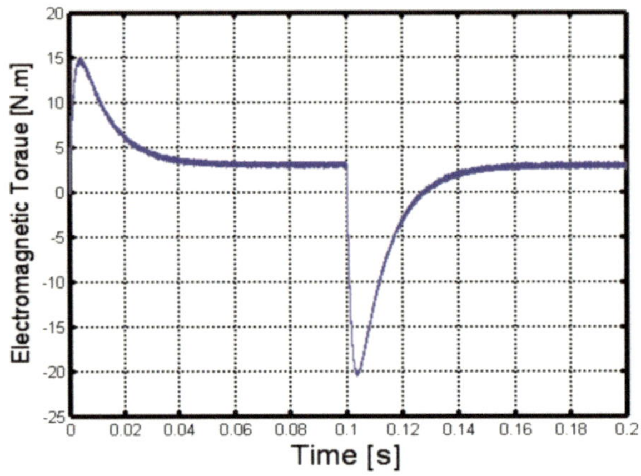

Fig. 12. Estimated electromagnetic torque.

The estimation algorithm is robust because the variation of the speed is important and the estimated speed follows the real speed when the motor starts and at the moment of speed inversion.

Figure 13. Presents trajectory of the estimated stator flux components, the stator flux trajectory is almost circular

Figure 14. Shows the rotor position.

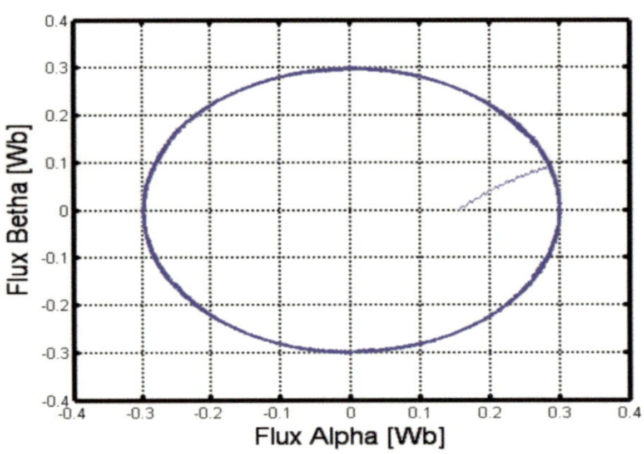

Fig. 13. Trajectory of the estimated stator flux

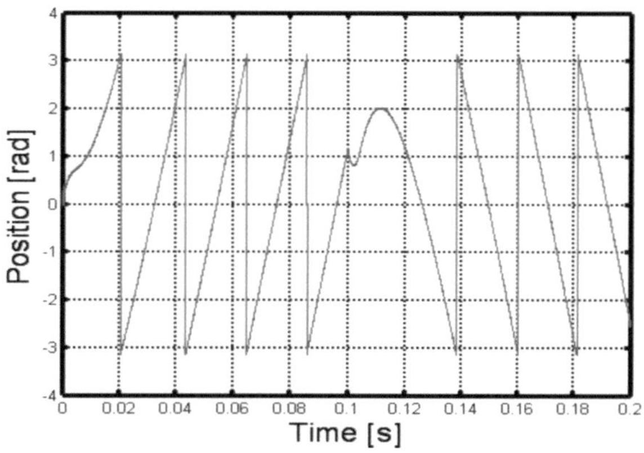

Fig. 14. Evolution of rotor position

7. Conclusion

In this paper, the EKF approach for a speed sensorless direct torque control (DTC) is proposed to estimate the rotor position and speed. The performance of the algorithm was investigated over a wide range of speeds. Overall, the simulation results obtained for estimating the speed and position are very satisfactory viewpoint estimation error, strength and stability of global training system in any operating condition.

8. References

Kawata, T.; Kosaka, M. & H. Uda. (2003). *Sensorless IPMSM drive with EKF estimation of speed and rotor position*, SICE Annual Conference in Fukui

Mademlis, C. & Agelidis, G. (2001). On *Considering Magnetic Saturation with Maximum Torque to Current Control in Interior Permanent Magnet Synchronous Motor Drives*, IEEE Trans. Energy Conversion, vol. 16, no. 3, Sep..

Thomas J.; René K. & Boel J. (2006). *Direct Torque Control of Permanent Magnet Synchronous Motors - An Overview*. 3rd IEEE benelux young researchers symposium in electrical power engineering, GHENT, BELGIUM

Kim Y. & Kook Y. (1999). *High Performance IPMSM Drives without Rotational Position Sensors Using Reduced-Order EKF*. IEEE Transactions on Energy Conversion, Vol. 14

Shi, K.; Chan, T.; Wong, Y. & Ho, S. (2000). *Speed estimation of an induction motor drive using extended Kalman filter*. proceeding IEEE

Pragasan, P. & Krishnan. R. (1988). *Modeling of permanent magnet motor drives*. IEEE Trans. Industrial electronics, vol. 35,

Tang, L.; Zhong, L.; Rahman, M. F. & Hu, Y.(2002) .*A Novel Direct Torque Control for Interior Permanent Magnet Synchronous Machine Drive System with Low Ripple in Torque and Flux-A Speed Sensorless Approach*. IEEE Trans Industry Applications Conference, 2002. 37th IAS Annual Meeting. Volume 1, Issue

Rahman, M. F.; Zhong, L.; Haque, E. & Rahman, M. A. (2003). *A Direct Torque-Controlled Interior Permanent-Magnet Synchronous Motor Drive Without a Speed Sensor.* IEEE Trans. Energy conversion, vol. 18, no. 1,

Bhavinkumar, S. (2004). *Field oriented control of step motors.* Thesis Master of Science. Dept. Elect. Eng Cleveland State Univ. India.

Velázquez, S. C.; Palomares, R. A. & Segura, A. N. (2004). *Speed Estimation for an Induction Motor Using the Extended Kalman Filter.* Proceedings of the 14th International Conference on Electronics, Communications and Computers

Yahia, K. ; Menacer, A.; Benakcha, A. ; Benchabane, F. ; & Taibi , D. (2005). *Estimation de la position et de la vitesse du moteur synchrone à aimants permanents en utilisant le filtre de Kalman étendu*. First International Conference on Electrical Systems. O. E. Bouaghi Univ. Algeria

Part 6

Switched Reluctance Motor Torque Control

6

Advanced Torque Control Scheme for the High Speed Switched Reluctance Motor

Dong-Hee Lee, So-Yeon Ahn and Jin-Woo Ahn
Dept. of Mechatronics Engineering, Kyungsung University
Republic of Korea

1. Introduction

High speed drive systems are much interested in the industrial application such as blower, compressor, pump and spindle due to the compact size and high efficiency. In recent, the demands of high speed drives are much increased due to the mechanical advantages of high speed system. SRMs (Switched Reluctance Motors) have simple structure and inherent mechanical strength without rotor winding and permanent magnet. These mechanical structures are suitable for harsh environments such as high temperature and high speed applications. Although SRMs have many advantages for the high speed applications, high torque ripple is still main problem to be applied to a high speed drives.

SRM(Switched Reluctance Motor) has been researched in last 150 years. However, significant amount of attention to this motor type was given in last few decades thank to the development of microcontrollers, power semiconductors and CAD technology. But compared with other motor, SRM cannot be operated without inverter. Therefore, it is suitable for variable speed application because inverter is essential.

SRM is a doubly salient machine because both the stator and rotor have salient poles. Flux is created by concentrated stator winding only. There are no permanent magnets, commutators, and windings in rotor side. So, SRM has simple and robust rotor structure that is very important in high speed application because centrifugal force is very high when rotor rotates at high speed. Due to a simple structure at the stator and the rotor, SRM can compete with other types of motor. And because SRM has only the windings at the stator, SRM achieves high efficiency especially in high speed region. With these advantageous features, SRM deserves to be investigated and it is good candidate for high speed drive systems in particular.

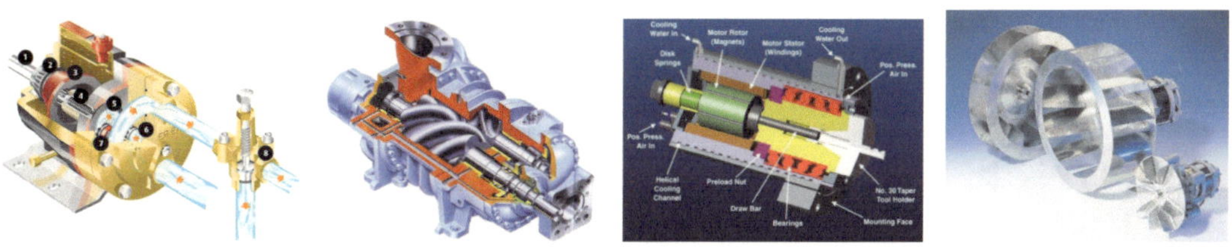

Fig. 1. Applications of a high speed drive.

2. Switched reluctance motor (SRM)

The SRM is an electric machine that converts the reluctance torque into mechanical power. In the SRM, both the stator and rotor have a structure of salient-pole, which contributes to produce a high output torque. The torque is produces by the alignment tendency of poles. The rotor will shift to a position where reluctance is to be minimized and thus the inductance of the excited winding is maximized. The SRM has a doubly salient structure, but there are no windings or permanent magnets on the rotor. The rotor is basically a piece of steel (and laminations) shaped to form salient poles. So it is the only motor type with salient poles in both the rotor and stator. As a result of its inherent simplicity, the SRM promises a reliable and a low-cost variable-speed drive and will undoubtedly take the place of many drives now using the cage induction, PM and DC machines in the short future. The number of poles on the SRM's stator is usually unequal to the number of the rotor to avoid the possibility of the rotor being in a state where it cannot produce initial torque, which occurs when all the rotor poles are aligned with the stator poles.

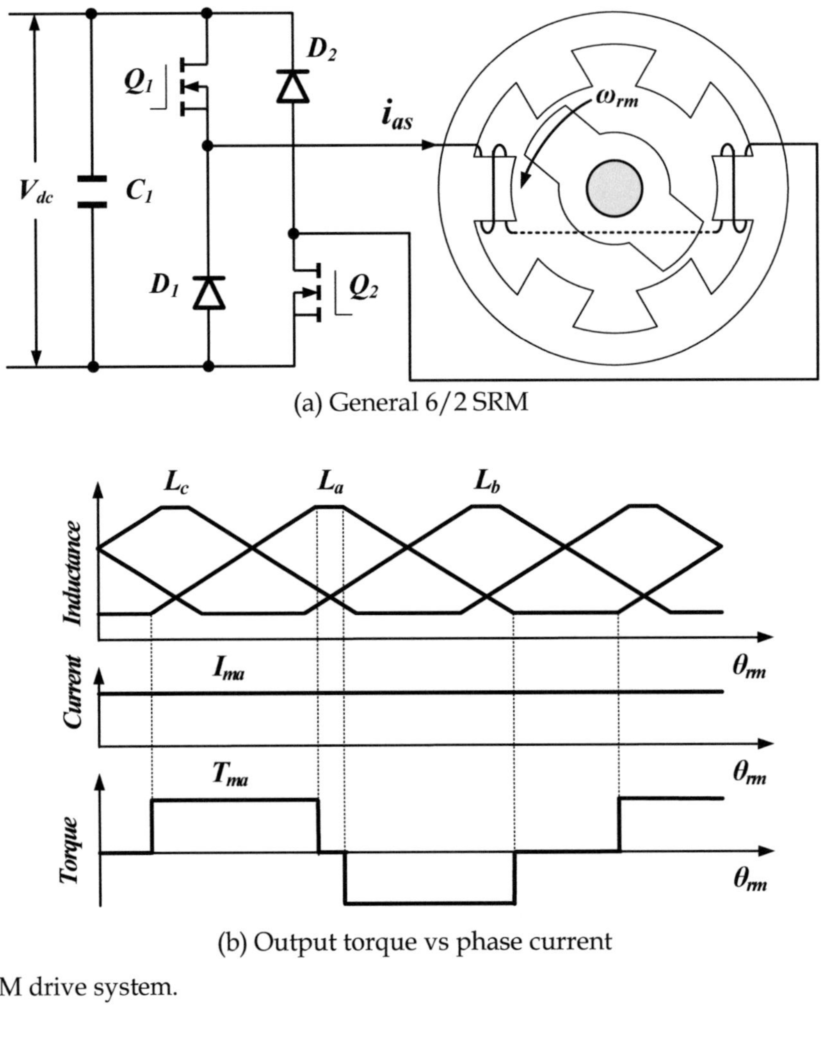

(a) General 6/2 SRM

(b) Output torque vs phase current

Fig. 2. SRM drive system.

Fig. 2 shows a general SRM drive system. With a constant phase current, the ideal phase torque is produced according to the square of current and inductance slope of the motor. Two basic equations of an SRM can be derived in terms of phase voltage and torque as follows.

$$v = R \cdot i + L(\theta_{rm}, i) \cdot \frac{di}{dt} + i \cdot \frac{dL(\theta_{rm}, i)}{d\theta_{rm}} \cdot \omega_{rm} \tag{1}$$

$$T_m = \frac{1}{2} \cdot i^2 \cdot \frac{dL(\theta_{rm}, i)}{d\theta_{rm}} \bigg|_{i=constant} \tag{2}$$

where, R: resistor of phase winding. θ_{rm} : rotor position, ω_{rm}: rotor speed, $L(\theta_{rm}, i)$: inductance is linearly varying with rotor position for a given current. However, the torque characteristics of SRM can be improved by the rotor and stator design.
Fig. 3 shows the various type of 4/2 SRMs. The modified types which have a staggered gap rotor pole surface, air teeth and air hole types of rotor pole are introduced.

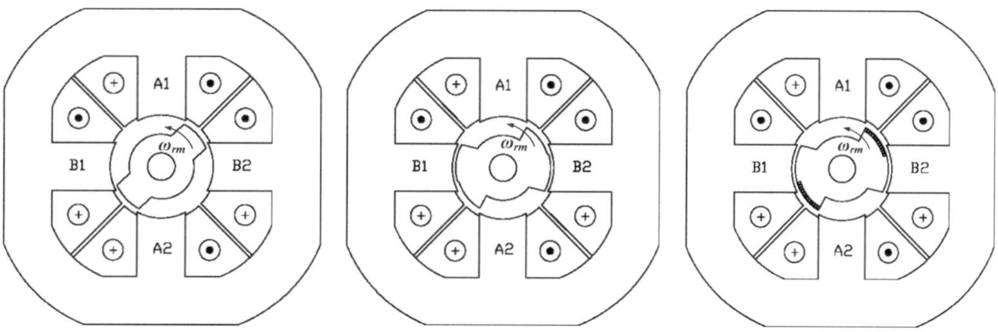

(a) Conventional type (b) Staggered gap type (c) Air teeth type

Fig. 3. Various type of 4/2 SRM.

Due to magnetic nonlinearity in an SRM, the phase inductance is nonlinear with respect to rotor position and phase current, and hence, constant torque profiling for torque ripple reduction is difficult compared to conventional AC motors such as PM and induction motors.

2.1 Mechanical structures

Constructions of SRM with no magnets or windings on the rotor also bring some disadvantage in SRM. Since there is only a single excitation source and but also magnetic saturation, the power density of reluctance motor is lower than PM motor. The construction of SRM is shown in Fig. 4. The dependence on magnetic saturation for torque production, coupled with the effects of fringing fields, and the classical fundamental square wave excitation result in nonlinear control characteristics for the reluctance motor. The double saliency construction and the discrete nature of torque production by the independent phases lead to higher torque ripple compared with other machines. The higher torque ripple, and the need to recover some energy from the magnetic flux, also cause the ripple current in the DC supply to be quite large, necessitating a large filter capacitor. The doubly salient structure of the SRM also causes higher acoustic noise compared with other machines. The main source of acoustic noise could induce the radial magnetic. So higher torque ripple and acoustic noise are the most critical disadvantages of the SRM.

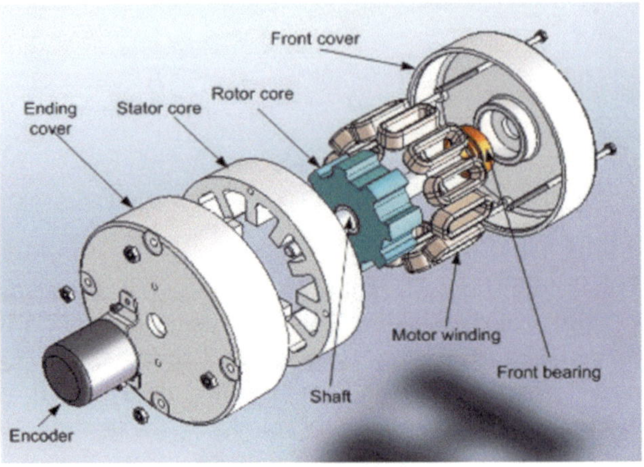

Fig. 4. The construction of SRM.

The absence of permanent magnets imposes the burden of excitation on the stator windings and converter, which increases the converter kVA requirement. Compared with PM brushless machines, the per unit stator copper losses will be higher, reducing the efficiency and torque per ampere. However, the maximum speed at constant power is not limited by the fixed magnet flux as in the PM machine, and hence, an extended constant power region of operation is possible in SRM.

2.2 Torque and torque ripple

The torque ripple of the SRM is divided to three parts. The first source is from the inherent torque ripple due to the mechanical and magnetic structure. Single-and two-phase SRMs have inherent torque dead-band due to an absence of the torque over-lap region between phases. Fig. 5 and Fig. 6 show the output torque according to the phase current in conventional single-and two-phase SRMs. As shown in Fig. 5 and 6, the positive torque region has dead-band from the inherent magnetic structure.

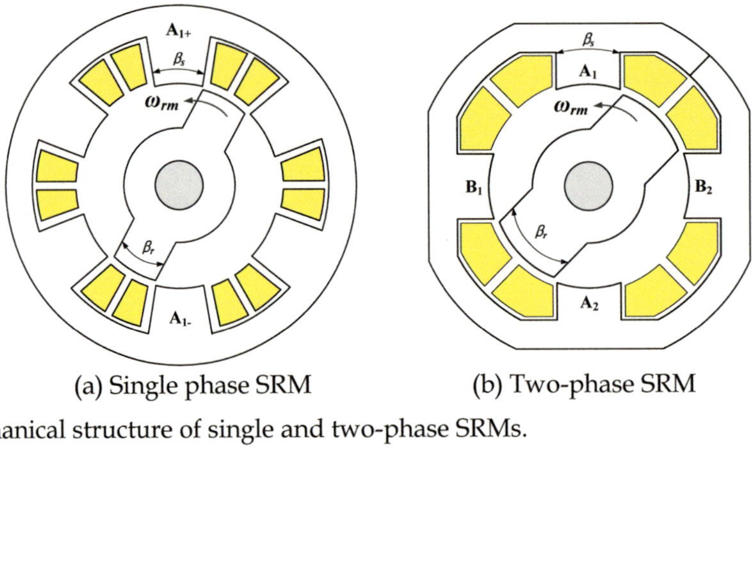

(a) Single phase SRM (b) Two-phase SRM

Fig. 5. Mechanical structure of single and two-phase SRMs.

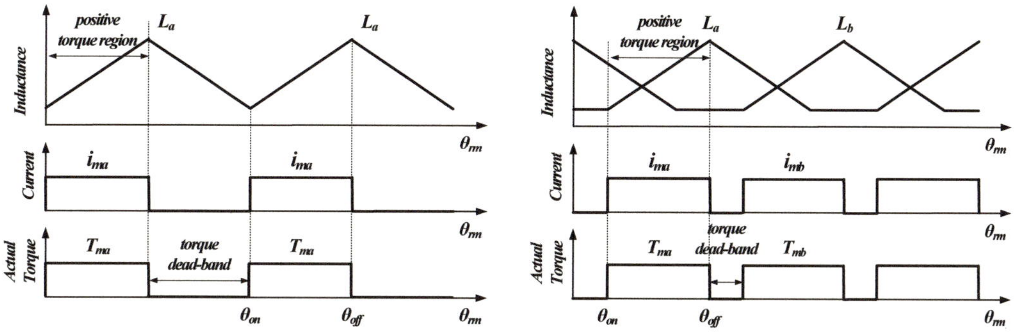

(a) Single phase SRM (b) Two-phase SRM

Fig. 6. Inductance, current and output torque of single and two-phase SRMs.

In order to overcome the inherent torque dead-band, asymmetric SRMs are investigated. These types have wide positive torque region with a short negative torque region, and are useful in an unidirectional application such as blowers and fans. For the single phase SRM, hybrid type using permanent magnet is researched. The inserted permanent magnet can produce cogging torque during torque dead-band. However, the torque of the inserted PM poles cannot be controlled, so it has high torque ripple but smaller than a conventional single-phase SRM. In two-phase SRM, rotor pole shapes can be changed to extend the torque region, and the inductance has asymmetric characteristics shown in Fig. 7.

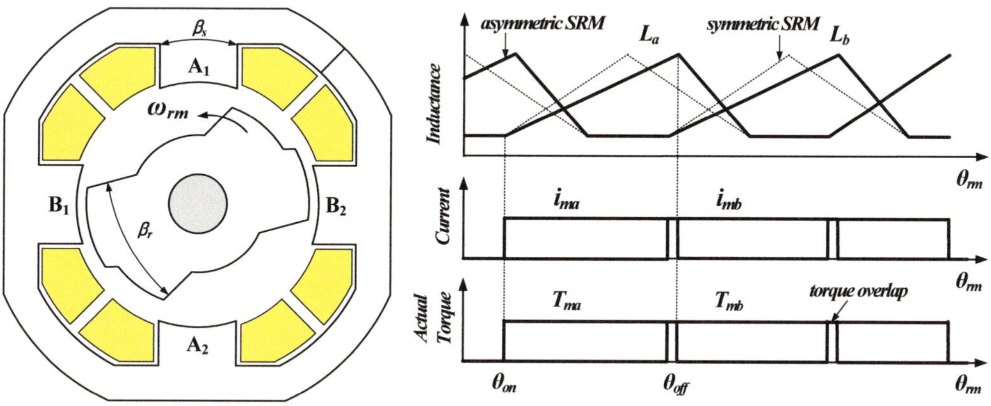

(a) Stepper rotor type SRM (b) Inductance, current and torque characteristics

Fig. 7. Uni-directional 4/2 SRM with torque overlap.

As shown in Fig. 7, the stepper rotor type has a wide rotor pole arc that produces torque overlap between two phases. In order to extend positive torque region, the inductance has asymmetric characteristics shown in Fig. 7(b), and is useful for an unidirectional application. It can remove torque dead-band, but torque ripple during stepper region is much serious. To reduce the torque ripple of stepper region, optimization design process can be adopted. From these research results, the inherent torque dead-band can be reduced.

The other is from the non-linear characteristics of inductance and un-constant torque by the constant current due to the saturation effect. Although the output torque is proportional to the inductance gradient, the inductance gradient is not constant. The inductance has non-linear according to rotor position due to the saturation effect. So, the constant phase current cannot guarantee a constant torque shown as Fig. 8. In the compact size motor, flux saturation is essential. So, the output torque ripple is different according to the load condition.

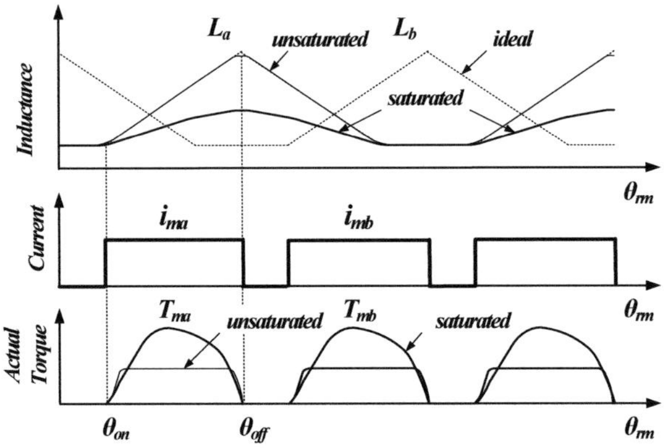

Fig. 8. Non-linear torque from the saturation effect.

In order to reduce the torque ripple from the saturation and non-linear characteristics, advance torque control scheme is required. In the next chapter, the detailed torque control schemes will be explained.

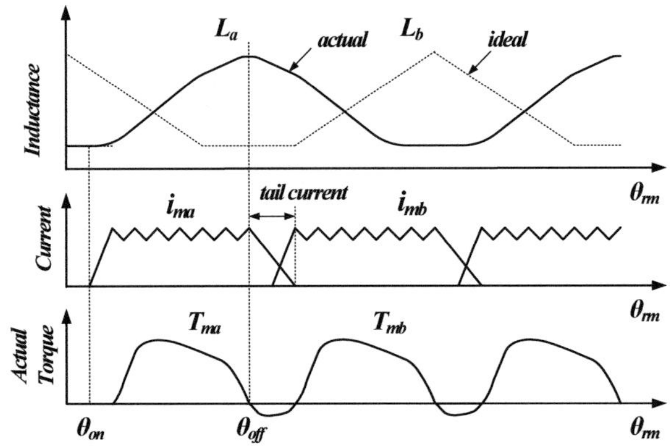

Fig. 9. Torque ripple from the tail current.

One reason of torque ripple is negative torque due to the tail current. The output torque of a phase is changed by the inductance slope. If the phase current is extended to the opposite torque region, the current produces the opposite torque. In the high speed or heavy load condition, the phase current is not extinguished in the same torque region, and the extended tail current produces negative torque as shown in Fig. 9. When the turn-off angle is controlled to remove the tail current, negative torque can be removed. However, the tail current is much serious in the high speed application due to a short commutation time. In order to reduce the negative torque effect, additional torque compensation algorithm and boost converter with a high negative demagnetization voltage have been researched. If the turn-off angle is much advanced to reduce the tail current, the effective torque region is much decreased with low efficiency and low output torque. The boost converter which can supply the high excitation and demagnetization voltage can reduce the commutation time, and it is useful to remove the tail current.

3. Torque control schemes

Various torque control schemes are investigated for the efficiency and torque ripple reduction. The torque control of SRMs is classified by three categories : on/off angle control, current control and direct torque control. The on/off angle control and current control schemes are simple and don't use any torque estimation. For a better performance, current control scheme uses on/off angle changing according to the load condition. In these control schemes, they didn't consider non-linear characteristics. So, the output torque of SRM is dependent on the saturation effect and inductance characteristics.

Fig. 10, 11 and 12 show the phase current and output torque of on/off angle control, current control and current control with on/off angle controller. The on/off angle control scheme uses a single pulse voltage during turn-on and-off angle (dwell angle). As shown in Fig. 10(b), the output torque is controlled by average value during dwell angle , and PWM switching is not used. So, the switching loss is very small compared with the current control scheme, but the output torque has large ripple. In the Fig. 10(a), the torque of phase A is controlled by turn-on angle controller, and phase B is controlled by turn-off angle controller. When the turn-on angle is advanced ($\theta_{onA1} < \theta_{onA3}$), the phase current and output torque is changed according to the turned on region. Much increased phase current i_{ma1} can increase the output torque than the small phase current i_{ma3} . Similarly, when the turn-off angle is delayed ($\theta_{offB1} < \theta_{offB3}$), the phase current and output torque range is changed according to the turned off angle. Much extended phase current i_{mb3} can increase the output torque than the short phase current i_{mb1} . This control scheme just controls average torque, and cannot control the instantaneous torque.

Fig. 11 shows a current control block diagram, phase current and output torque. Hysteresis current controller or PWM(Pulse Width Modulation) can be used to adjust the phase current. In the current controller, the activated phase is determined by the rotor position. By the switching of power devices, the switching loss is much increased than the on/off angle control method. But, excitation current can be controlled with flat-top shape. If the output torque is linearly proportional to the phase current, a constant output torque can be obtained. However, the output torque of SRM has non-linear characteristics due to the inductance and saturation effect. So, the output torque is dependent on the non-linear torque characteristics.

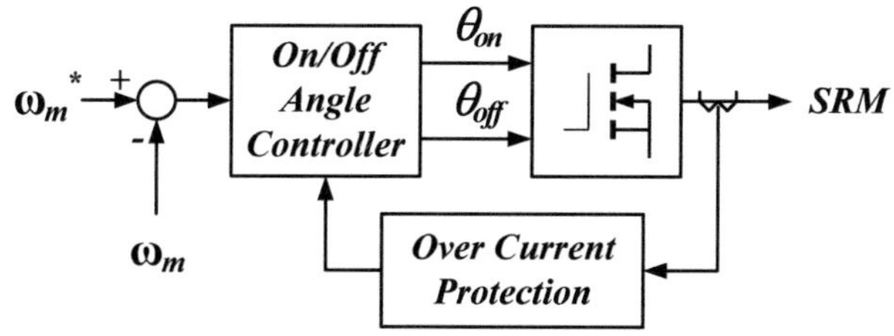

Fig. 10. (a) On/off angle controller.

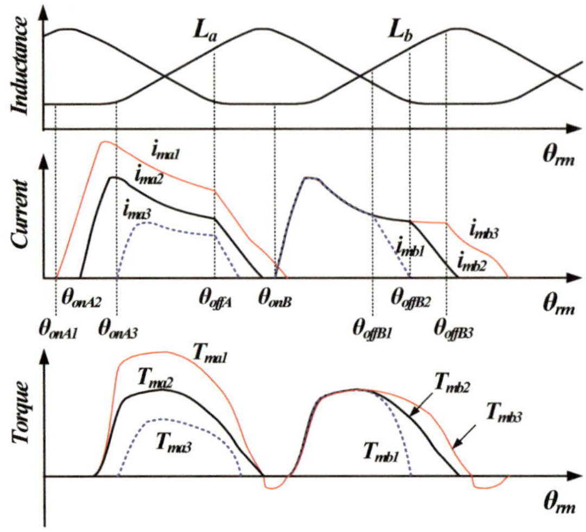

(b) Phase current and output torque.

Fig. 10. Phase current and output torque of the on/off angle control scheme.

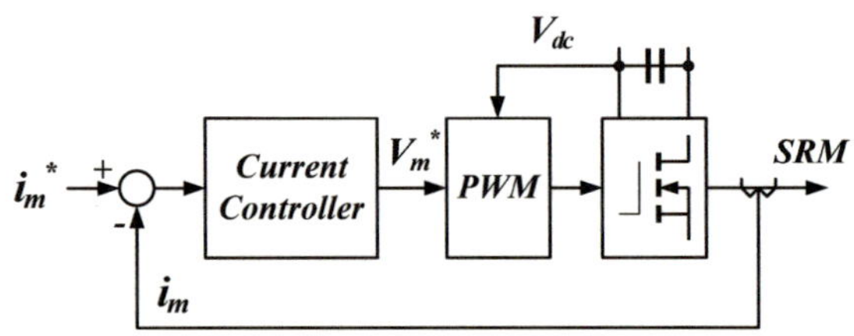

(a) Current control scheme

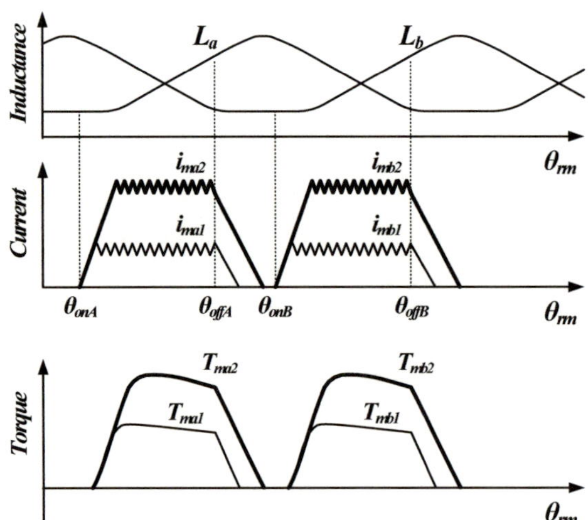

(b) Phase current and output torque

Fig. 11. Phase current and output torque of current control scheme.

In order to build up the desired current and extinguish the demagnetized current, turn-on and-off angles are very important. So, the current controller with angle adjusting can increase the torque performance. Although, the output torque is dependent on the non-linear characteristics, but the torque region can be adjusted according to the load condition. In a heavy load with high reference current, the turn-on angle is advanced to ensure the excitation current building-up time. On the contrary, the turn-off angle is delayed in the light load with low reference current due to the efficiency. Because the output torque cannot be produced during minimum inductance and maximum inductance region, the phase current of these region is much contributed to the loss.

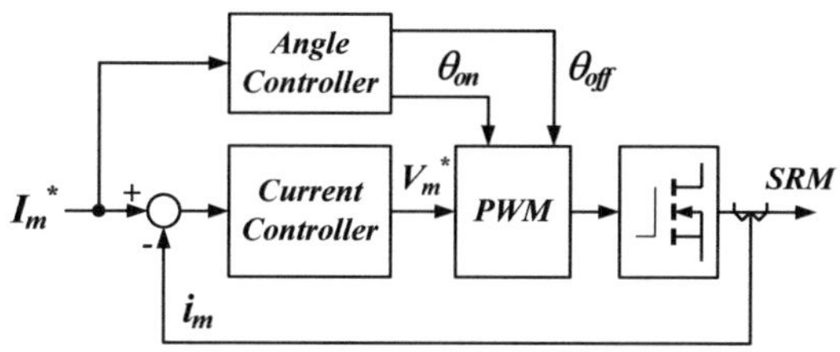

(a) Current control scheme with angle controller

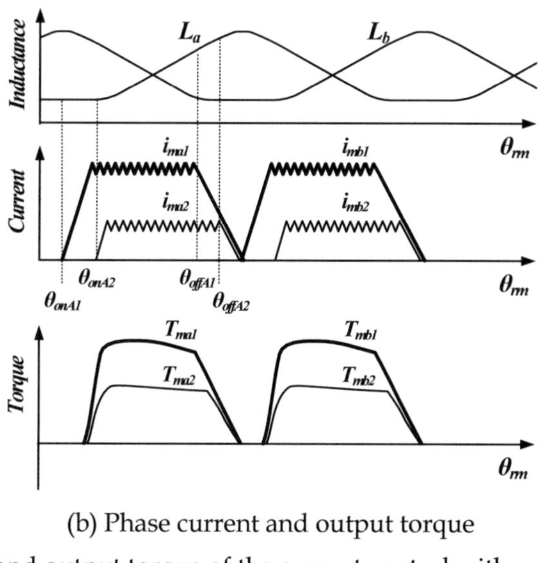

(b) Phase current and output torque

Fig. 12. Phase current and output torque of the current control with on/off angle controller

The other torque control methods are current shape, direct torque control and TSF(Torque Sharing Function) technologies. These torque control methods are complex but torque ripple can be reduced. In order to reduce the torque ripple, the non-linear characteristics are considered in these control schemes. To compensate the output torque variation according to the current and rotor position, motor characteristics are included in the controller by look-up table, data memory and simplified mathematical model.

The current shape control method uses a compensation current according to the rotor position to compensate the saturation effect. In this control method, the compensation term is calculated with a simplified model. So, the torque ripple is not perfectly rejected.

More advance torque control method uses a torque to current data to reduce the torque ripple. In this approach, the reference current is changed according to the rotor position and reference torque with the motor characteristics. The torque to current data is determined by the actual motor characteristics and has 3-dimensions. Fig. 13 shows the torque control scheme using torque-to-current data, and its operating characteristics consideration. As shown in Fig. 13(b), the reference current to produce the desired output torque is not constant, and non-linear current reference is determined by the torque-to-current data. The current controller is designed to keep the reference current. This approach needs a complex database to compensate the motor characteristics, but the torque ripple can be reduced. The output torque is dependent on the accuracy of the torque-to-current data and current controller. When the torque-to-current data is very accurate, the output torque is very constant during a single phase excitation. But, it has torque ripple during the commutation region.

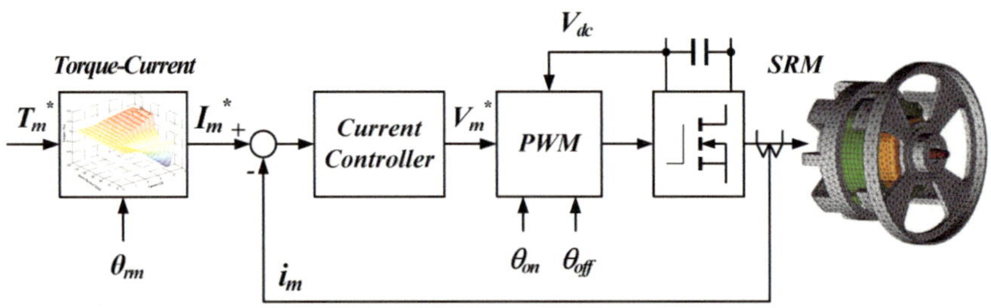

(a) Non-linear current control scheme

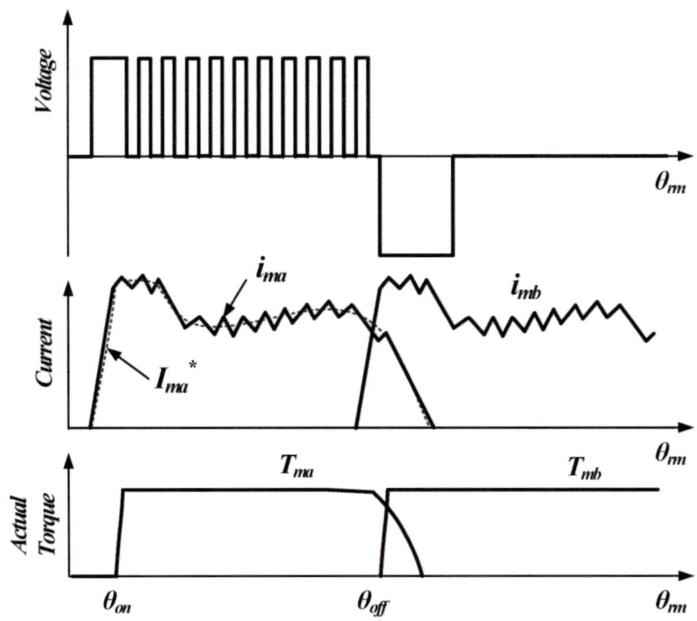

(b) Operating characteristics

Fig. 13. Non-linear current control scheme and its operating characteristics.

Another approach such as TSF uses a torque references and torque sharing during commutation region. In the TSF method, torque region is divided by two region : single-phase mode and two-phase mode. In a single-phase mode, only one phase current produces the output torque. In a two-phase mode, the outgoing phase and incoming phase current produce the output torque, and the total torque is the sum of the two phases. In order to reduce the torque ripple during the commutation region, torque references should be changed, and the sum of the torque references of the each phases are constant. The reference torque of the outgoing phase is decreased, and the reference torque of the incoming phase is increased during this region. The increasing and decreasing torque reference can be determined as linear and sinusoidal function. Fig. 14 shows a TSF method using sinusoidal function.

As shown in Fig. 14(b), torque reference of phase A is increased, and reached the target value between θ_{onA} and θ_{offB}, the reference of phase B is decreased from target value to zero, reversely. However, the sum of two-phase is same as target value. When the phase current can keep the reference current, the output torque of SRM can keep the constant value with a small torque ripple due to the current switching. Similar to the non-linear current control scheme, TSF needs a torque-to-current data to produce the reference current which generates a reference torque according to the rotor position. As we shown in previous control scheme, the output torque of the TSF method is dependent on the accuracy of the data and current control performance.

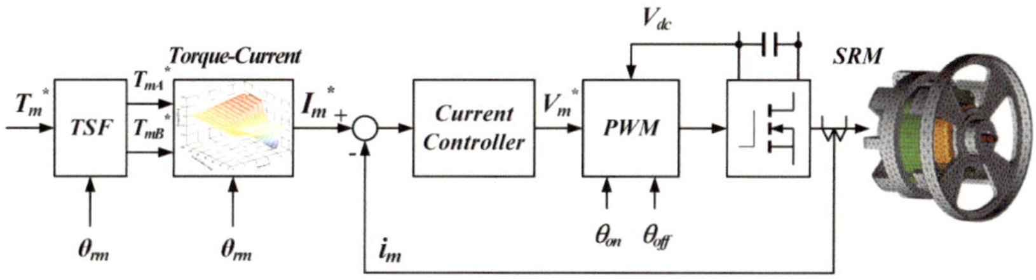

(a) TSF control scheme

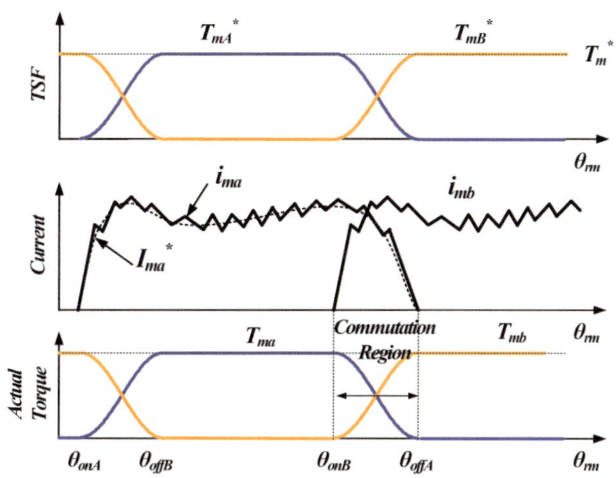

(b) Operating characteristics of the TSF control scheme

Fig. 14. TSF control scheme and its operating characteristics.

The torque reference during one-phase mode of the each phases is defined as follows.

$$f^*_{T(k)} = T^*_m$$
$$f^*_{T(k-1)} = 0 \tag{3}$$

And the torque commands during commutation are defined as follows in the cosine TSF.

$$f^*_{T(k)} = \left[1 - \cos\left(\frac{\theta_{rm} - \theta_{on(k)}}{\theta_{off(k-1)} - \theta_{on(k)}} \cdot \frac{\pi}{2}\right)\right] T^*_m$$
$$f^*_{T(k-1)} = 1 - f^*_{T(k)} \tag{4}$$

Where, θ_{rm} is rotor position. The $\theta_{on(k)}$ and $\theta_{off(k-1)}$ are turn-on and-off angles of the incoming and outgoing phases, respectively.
Compared with TSF method, DTC(Direct Torque Control) scheme uses a torque estimator and simple switching rules to control the output torque. The torque estimator is similarly made using the non-linear torque characteristics of the motor according to the phase current and rotor position. And these data reflect the saturation effect and the inductance characteristics. The DTC uses a direct torque comparison method with the torque estimator, the switching rules are designed to compensate the torque error between the torque reference and the estimated torque. The switching technologies of the DTC algorithm act as hysteresis controller. For the effective compensation of the torque error, the switching technologies can use PWM method.

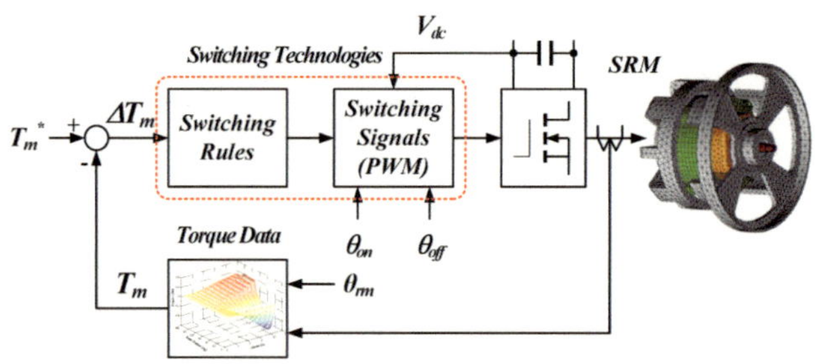

(a) Block diagram of the direct torque control scheme.

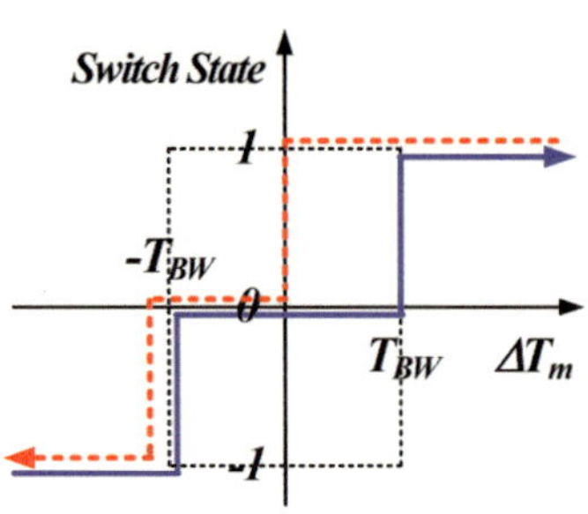

(b) Switching rule without PWM

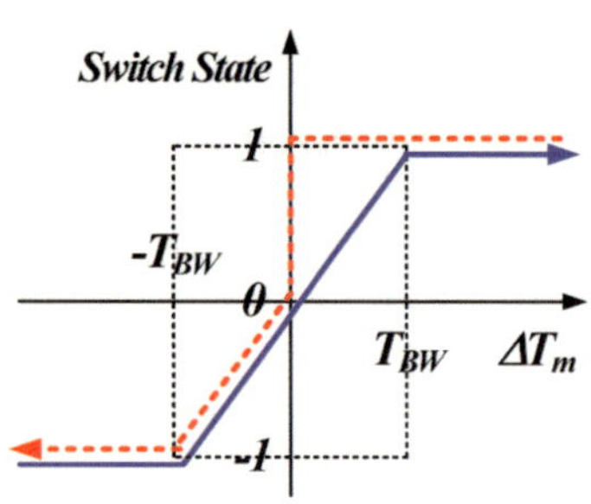

(c) Switching rule with PWM

Fig. 15. Direct torque control scheme.

Fig. 15 shows the DTC control block diagram and the examples of switching rules. The switching state of the switching rule means the operating modes of converter. And the T_{BW} is the torque error band of the DTC. The switching state 1 is excitation mode which supplies the dc-link voltage in the phase winding. The switching state 0 is free-wheeling mode which supplies zero voltage, and one power devices of the converter is turned on. And the switching state -1 is the demagnetization mode which supplies negative dc-link voltage in the phase winding. During the demagnetization mode, power devices of the phase winding are fully turned off. With the PWM technology, the switching state can move more smoothly. The PWM duty ratio is controlled by the torque error within the switching states. The PWM method can guarantee the fixed switching frequency with small torque ripple.

The previous researches are very useful to reduce the torque ripple of SRM in the conventional speed region. In the region, the main torque ripple is from the inherent torque dead-zone and non-linear torque characteristics. So, the DTC and TSF can reduce the torque ripple from the non-linear torque characteristics with the accurate torque data.

4. Advanced torque control scheme for a high speed SRM

4.1 Advanced converter topologies

When the motor speed is increased, the switching and excitation time are much decreased. The phase current of an AC machine has sinusoidal waveform, and current changing is very smooth. However, the phase current of SRM has square waveform. And the changing of the phase current is very rapid. Sometimes, the excitation current cannot be reached to the desired value, and the demagnetization cannot be extinguished during the commutation time due to a short excitation and demagnetization time in a high speed region. The insufficient excitation current makes an insufficient output torque, and the extended demagnetization can produce a negative torque. The much advanced turn-off angle can reduce the tail current in a high speed, but the output torque is not sufficient due to a short excitation region.

Fig. 16 shows the current and torque waveforms in a high speed region.

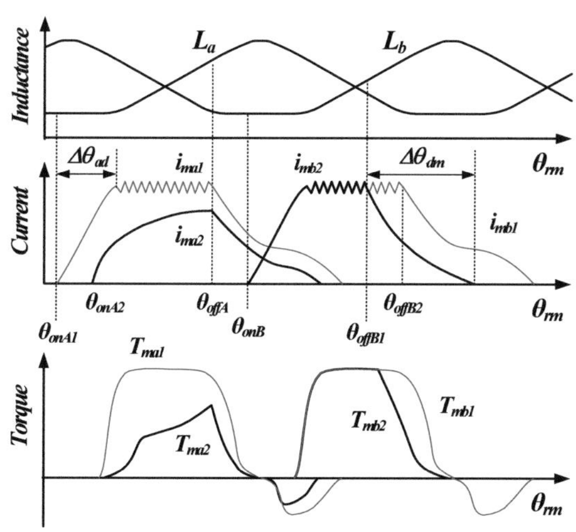

Fig. 16. Current and torque in a high speed.

As shown in Fig. 16, a short excitation and demagnetization time can produce a high torque ripple. Compared with i_{ma1} and i_{ma2}, the phase current i_{ma2} cannot be reach the desired value due to a short excitation time. So, the output torque has a high torque ripple. With the extended advance angle, the phase current can be reached to the desired value. Similarly, the turn-off angle should be advanced to reduce the tail current. In Fig. 16, the phase current i_{mb1} is extended to the negative torque region, and it produces a negative torque due to the tail current. With an advance turn-off angle, the phase current i_{mb2} is extinguished during the proper torque region, but the active torque region is decreased. The decreased active torque region from the advanced turn-off angle makes high torque ripple due to the torque dead-band.

Fig. 17 shows the phase current in the high speed region with an asymmetric converter. As shown in Fig. 17, the motor is excited at θ_{on} advanced $\Delta\theta_{ad}$ from the start point of positive torque region θ_{min} to establish sufficient torque current. The desired phase current is represented by dashed line in Fig. 17 and is demagnetized at θ_{off}, and decreased to zero before the starting point of the negative torque region θ_{max} so as not to develop negative torque.

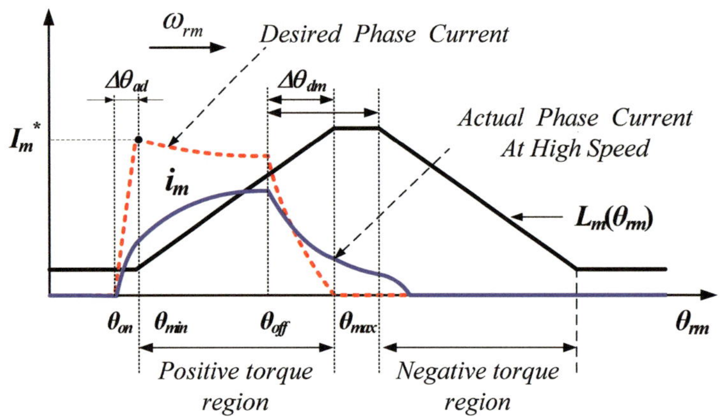

Fig. 17. Current in a high speed region.

In order to secure enough time to build-up the desired phase current I_m^*, the advance angle $\Delta\theta_{ad}$ can be adjusted according to motor speed ω_{rm}. From the voltage equations of the SRM, the proper advance angle can be calculated by the current rising time as follows regardless of phase resistance at the turn-on position

$$\Delta\theta_{ad} = L_{\min} \frac{I_m^*}{V_m} \tag{5}$$

$$\Delta\theta_{dm} = L_m(\theta_{off}, i_m) \frac{i_m}{V_m} \tag{6}$$

$$\Delta t_{ad} = \frac{\Delta\theta_{ad}}{\omega_{rm}} \tag{7}$$

$$\Delta t_{dm} = \frac{\Delta\theta_{dm}}{\omega_{rm}} \tag{8}$$

Where, I_m^* is the desired phase current, and V_m is the terminal voltage of the phase winding. As speed increases, the advance and demagnetization angles are increased and turn-of and turn-off positions may be advanced to the negative torque region of the previous phase.
If the actual terminal voltage is assumed as dc-link value, the actual maximum phase current can be obtained when the advance angle is in the previous unaligned position.

$$I_{\max} = \frac{\Delta\theta_{ad} \cdot V_{dc}}{L_{\min} \cdot \omega_{rm}} \tag{9}$$

And, the maximum output torque at θmin can be derived as follow.

$$T_{\max} = \frac{1}{2} \cdot I_{\max}^2 \frac{dL(\theta_{rm})}{d\theta_{rm}} \tag{10}$$

With a fixed turn-on position, the actual phase current which is denoted by a solid line cannot reach the desired value in the high speed region as shown in Fig. 17.
In a high speed, the advance and demagnetization time are much decreased. In order to reduce the torque ripple, the phase current should be well controlled. However, the short excitation and demagnetization times make a high current error in a high speed region.
In order to overcome these problems, boost voltage convertershas been researched. Fig. 18 shows the 4-level converter for high speed SRM.

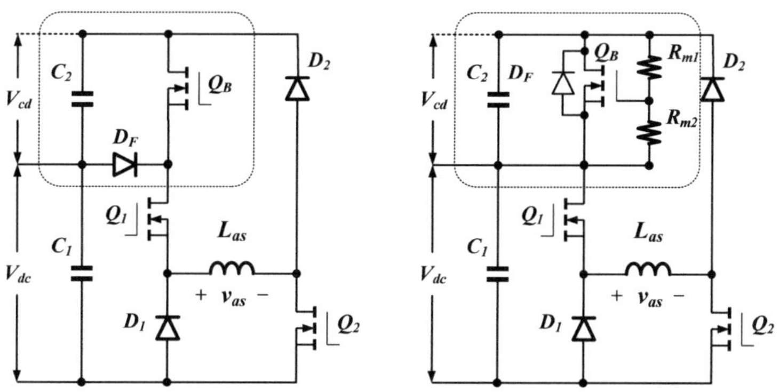

(a) 4-level type 1 (b) 4-level type 2

Fig. 18. 4-level converter for high speed drive.

The 4-level converter has an additional charge capacitor *C2*, a power switch *QB* and a diode *DF* compared with general asymmetric one. In the type 2, the power diode D_F is included as an anti-parallel diode of power switch Q_B. So, it can reduce the additional power devices to implement the 4-level converter. The charge capacitor C_2 recovers energy from the phase during demagnetization, and the phase current is quickly reduced due to the high negative bias. Then the charged high voltage is through the power switch Q_B to the next excitation phase winding for the fast phase current build-up in type 1. Differ from the type 1, the power switch of the type 2 is controlled to keep the boost voltage as a fixed value. And the excitation voltage is same as the conventional asymmetric converter.

Fig. 19, 20 and 21 show the operating modes of the 4-level type 1 and 4-level type 2. In the fast excitation mode(Fig. 19(a)), dc-link voltage V_{dc} and charged voltage V_{cd} are supplied to excite the phase winding, then the excited phase current builds up fast by the high positive voltage. Similarly, the demagnetized phase current during turn-off is decreased quickly by the high negative bias in the fast demagnetization mode(Fig. 19(d) and Fig. 20(d)). The excitation and wheeling modes are same as the conventional asymmetric converter. However, the demagnetization mode(Fig. 20(c)) supplies the negative boost voltage to the phase current.

The 4-level type 1 is very useful to the high speed drive due to the fast excitation and fast demagnetization modes. However, the boost voltage which is charged in C_2 should be controlled to keep the constant value. To prevent the higher charged voltage, the boost voltage is always adjusted, so the additional voltage sensor is required. In order to discharge the higher boost voltage, the fast excitation voltage can be supply in the positive torque region. So, the control scheme is very complex. Compare with type 1, the boost voltage of the type 2 is automatically adjusted by the gate resistor R_{m1} and R_{m2}. And the operating modes are always independent. This makes easy control of the SRM. And it can reduce the power devices to implement 4-level, although fast excitation is impossible. But, it supplies negative boost voltage during the demagnetization mode, and the current control is easier than the type 1.

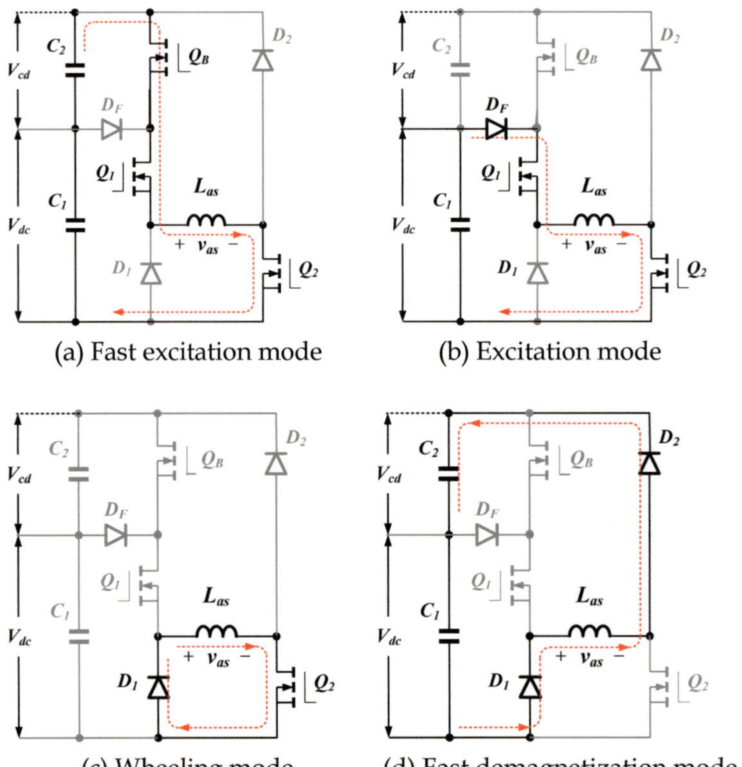

(a) Fast excitation mode (b) Excitation mode

(c) Wheeling mode (d) Fast demagnetization mode

Fig. 19. Operating modes of the 4-level converter (type 1).

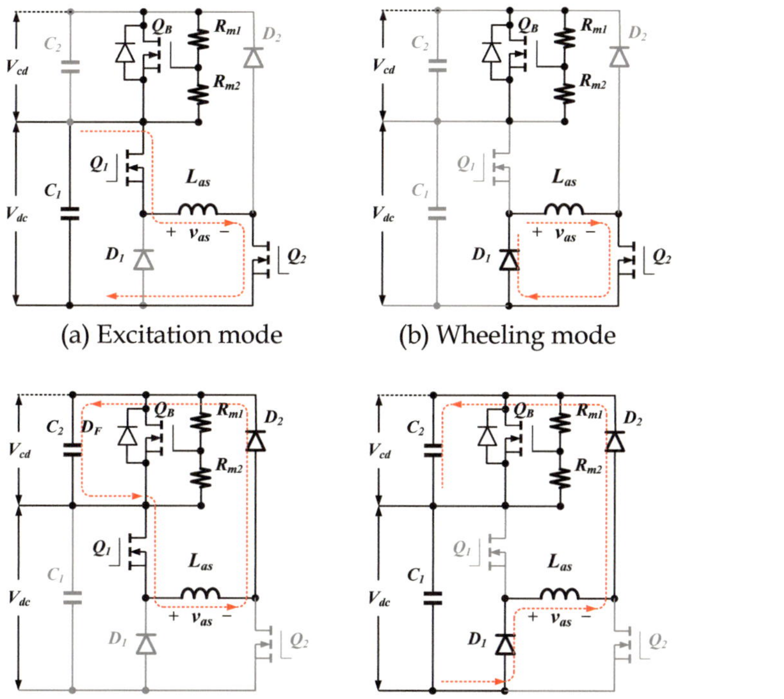

(a) Excitation mode (b) Wheeling mode

(c) Demagnetization mode (d) Fast demagnetization mode

Fig. 20. Operating modes of the 4-level converter (type 2).

With a PWM(Pulse Width Modulation) technology, two operating modes can be used in a sample time to control the phase current.

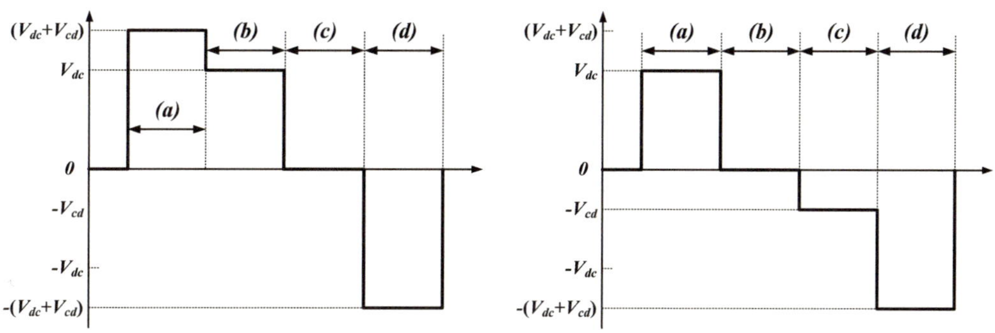

(a) Phase voltage of the type 1 (b) Phase voltage of the type 2

Fig. 21. Phase voltage of the 4-level converter.

Fig. 22 shows current waveform of the 4-levle converter type 1 and asymmetric one at 10,000[rpm] from experiment in a 12/8 SRM. As shown in Fig. 22, excitation current of the 4-level converter can support fast build-up and commutation of phase current when compared with a conventional asymmetric converter. Another approach using 4-level converter is DITC method. In the torque control, the additional control mode is very effective.

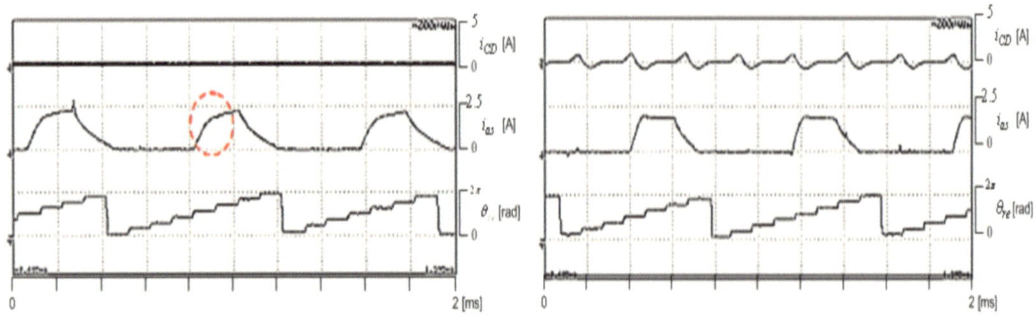

(a) Asymmetric converter (b) 4-level converter type 1

Fig. 22. Comparison of current waveform(10,000[rpm]).

Fig. 23 shows the proto-type high speed 4/2 SRM. The outer radius of the rotor is not constant to produce a wide and a constant positive torque. From this, it has asymmetric inductance which is wide positive and a short negative torque region.

(a) Rotor

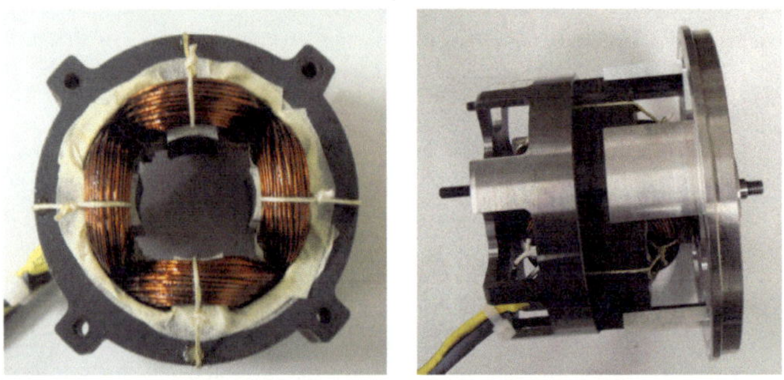

(b) Stator (c) Assembled SRM

Fig. 23. Proto-type high speed 4/2 SRM.

Fig. 24 and 25 show the simulation results of the 4-level type 2. The proto-type motor is 4/2 SRM which has asymmetric inductance. The proto-type SRM has wide positive torque region with over-lap between the phases, and designed for unidirectional applications.

Fig. 24 shows the compared simulation results with conventional asymmetric converter and the proposed 4-level type 2 at 30,000[rpm] with rated load 0.2[Nm] in the proto-type motor. As shown in Fig. 24, the tail current which is extended to the negative torque region produces an additional torque ripple. Compared with Fig. 24(a), the proposed control scheme shown in Fig. 24(b) can reduce the tail current and torque ripple due to the fast demagnetization with a high negative voltage level.

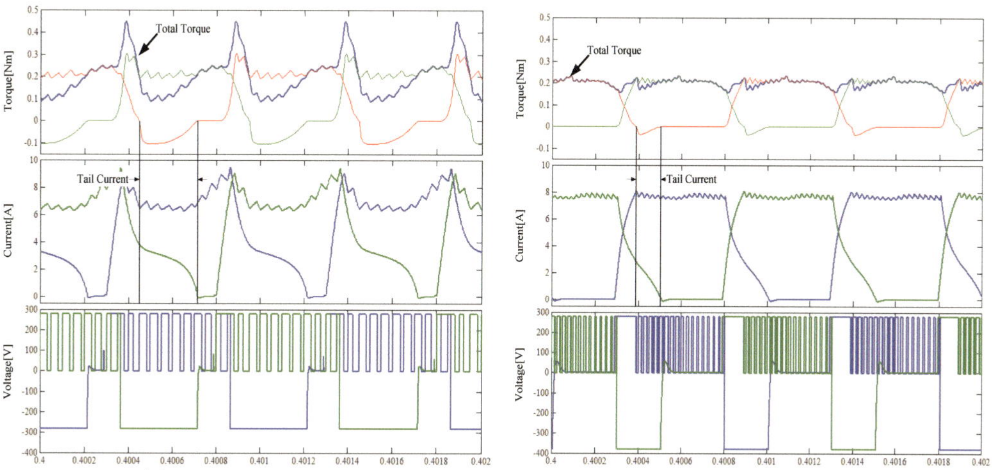

(a) Asymmetric converter (b) 4-level converter type 2

Fig. 24. Compared simulation results(30,000[rpm]).

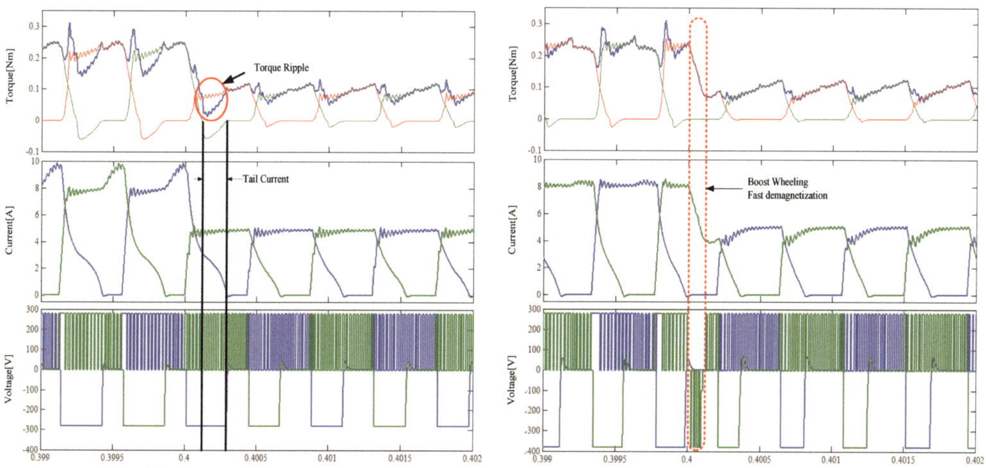

(a) Asymmetric converter (b) 4-level converter type 2

Fig. 25. Compared simulation results with reference variation.

When the reference current is sudden changed, an additional boost wheeling and fast demagnetization mode of the proposed 4-level type 2 can reduce the torque ripple and enhance the control dynamics shown in Fig. 25. Compare to Fig. 25(a), the proposed control scheme (Fig. 25(b)) shows less torque ripple next to the variation point as result of the boost wheeling and fast demagnetization.

4.2 Negative torque compensation

Another approach in the torque control for the high speed SRM, is torque compensation method. Fig. 26 shows the TSF algorithm in the general speed and the high speed region. As shown in Fig. 26(a) and Fig. 26(b), the each torque references are changed to the current reference to produce the reference torque with the considerations of non-linear torque characteristics. The controlled current can keep the reference torque in the conventional speed region shown in Fig. 26(a). However, the practical current cannot be kept the reference value due to the short commutation time in the high speed region. So, the tail current which is extended to the negative torque region, produces negative torque when the phase is inactive. From this tail current, the total torque of the high speed SRM has high torque ripple shown in Fig. 26(b).

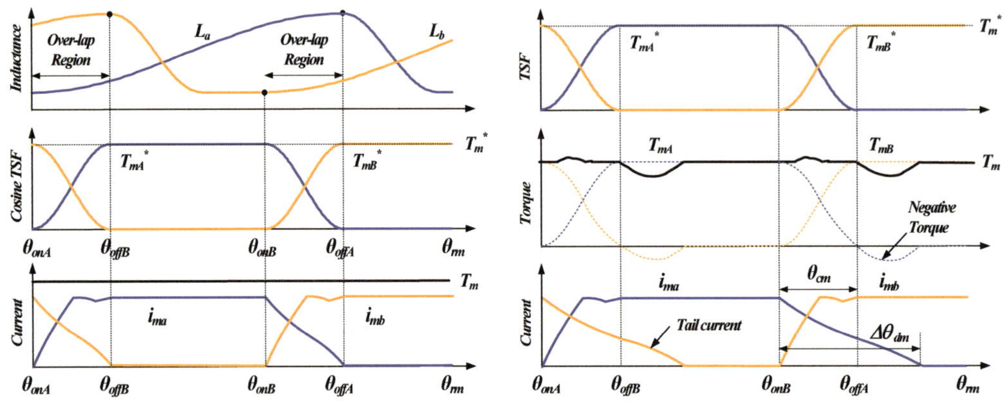

(a) Conventional speed region (b) High speed region

Fig. 26. Torque and current in TSF according to the speed.

In order to compensate the torque ripple from the tail current, the modified TSF control scheme is proposed. The proposed TSF has compensating torque block in the active phase. And the compensating torque can reduce the torque ripple from the negative torque of the inactive phase in a high speed. In order to reduce the tail current during commutation region, the switching signal of the outgoing phase is fully turned off. If the phase current is not extended to the negative torque region, the switching patterns of the each phases are determined by the torque errors during commutation region. However, the swtching signals of the outgoing phase are fully turned off, when the demagentization time is over than the designed commutation angle to reduce the negative torque from the tail current. And the torque error of the outgoing phase is compensated by the incoming phase current reference. After the commutation region, the negative torque of the outgoing phase is compensated by the compensation torque of the active phase.

Fig. 27 shows the block diagram of the proposed TSF control scheme for a high speed region. Compared with the conventional TSF method shown in Fig. 27(b), it has torque compensators and PWM limit signals.

In the Fig. 27, T_{cA}^* and T_{cB}^* are the compensating torques for the negative torques from the tail current. The compensating torque is simply calculated without any control gains as follows.

$$\begin{aligned} T_{cA}^* &= T_{mB}^* - T_{mb} \\ T_{cB}^* &= T_{mA}^* - T_{ma} \end{aligned} \tag{11}$$

Where, $T_{mA}^*$and T_{mB}^* are torque references of the each phases. And the T_{mA} and T_{mB} are the estimated torques which are calculated by the look-up table according to the actual phase current and rotor position. The EN_A and EN_B are the enable signals for the compensation controller.

The Current Reference block in Fig. 27(b) is designed as look-up table whose data are defined by the reference torque and rotor position with the non-linear characteristics of the proto-type SRM. Fig. 28 shows the torque estimator and current reference data which are used in this paper. The current controller is designed by the PI controller.

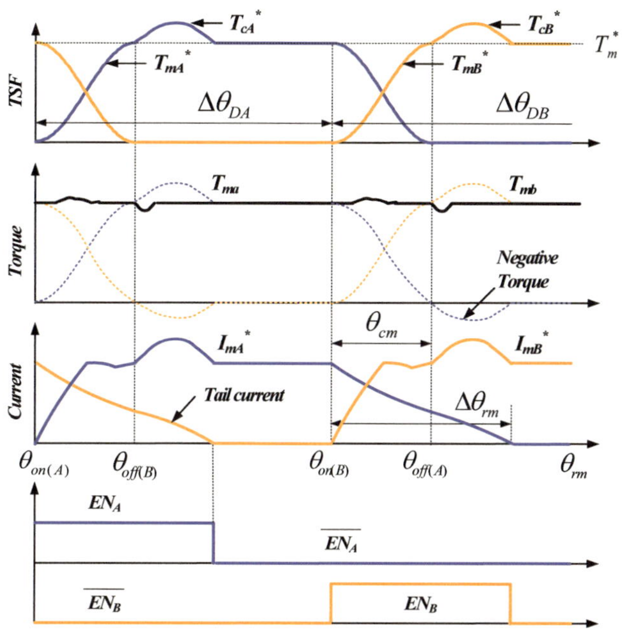

(a) Modified torque sharing function

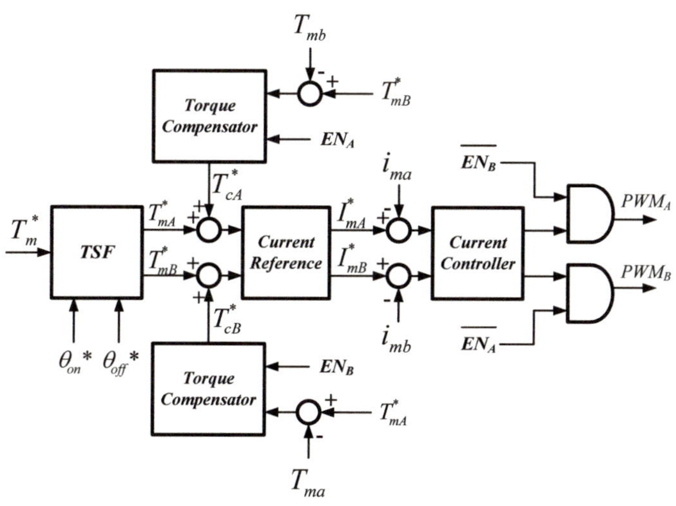

(b) Control block diagram

Fig. 27. Proposed modified TSF control scheme.

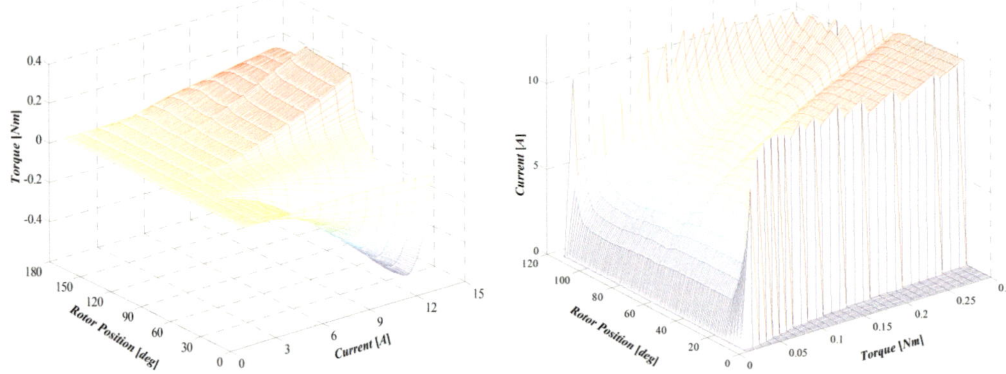

(a) Look-up table of the torque estimator (b) Look-up table of the current reference

Fig. 28. The torque estimator and current reference of the proto-type SRM.

Fig. 29 shows the compared simulation results in the conventional TSF and the proposed control scheme at 10,000[rpm]. As shown in Fig. 29, the simulation results of the proposed type are marginally improved, and the torque ripple is almost same due to the a small tail current. In the 10,000[rpm], the negative torque is very small and the compensation effect is not much. The output torque of the high speed SRM can be controlled well by the TSF control scheme.

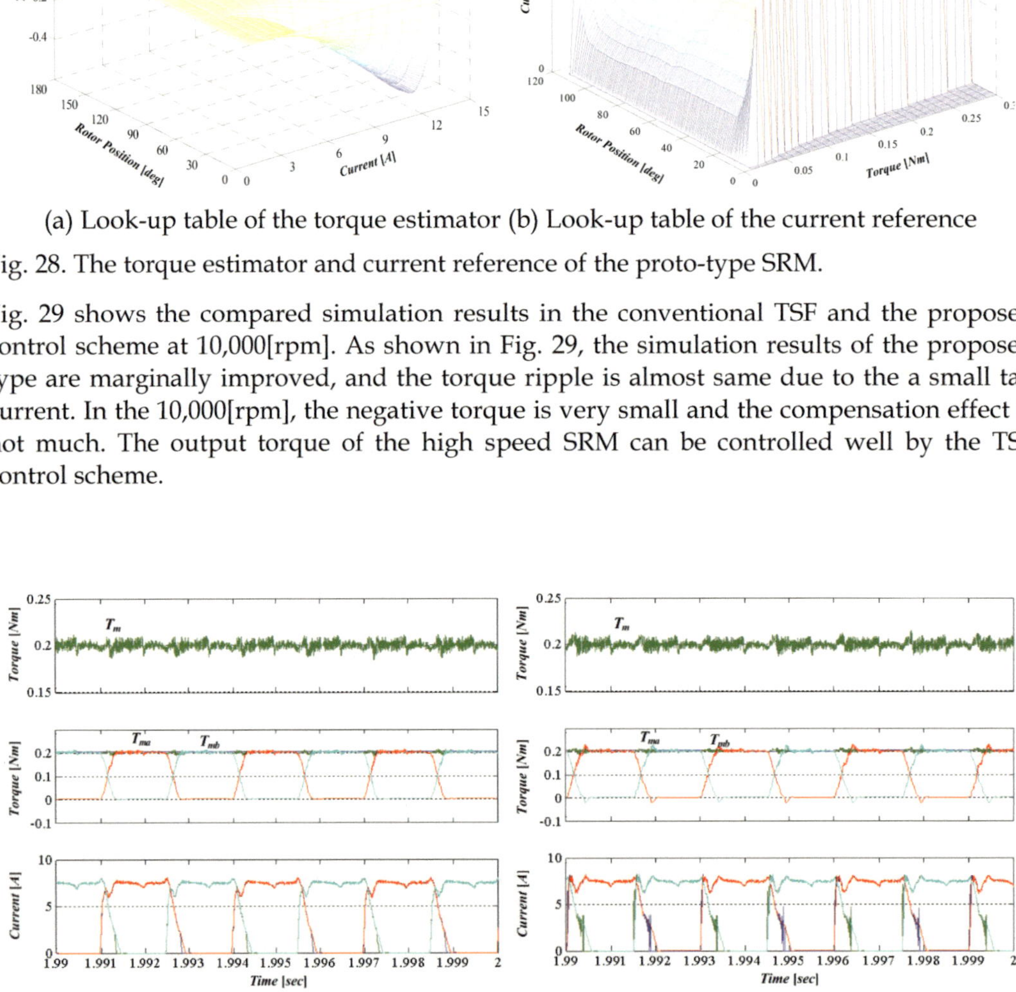

(a) Conventional cosine TSF (b) Proposed TSF control scheme

Fig. 29. Compared simulation result at 10,000[rpm].

Fig. 30 shows the compared simulation results at 30,000[rpm]. As shown in Fig. 30, the torque ripple of the proposed control scheme is much reduced than the conventional one. The torque ripple of the conventional TSF is very serious after the commutation region due to the negative torque from the tail current of the outgoing phase winding. However, the proposed control scheme can compensate the negative torque with the active phase torque.

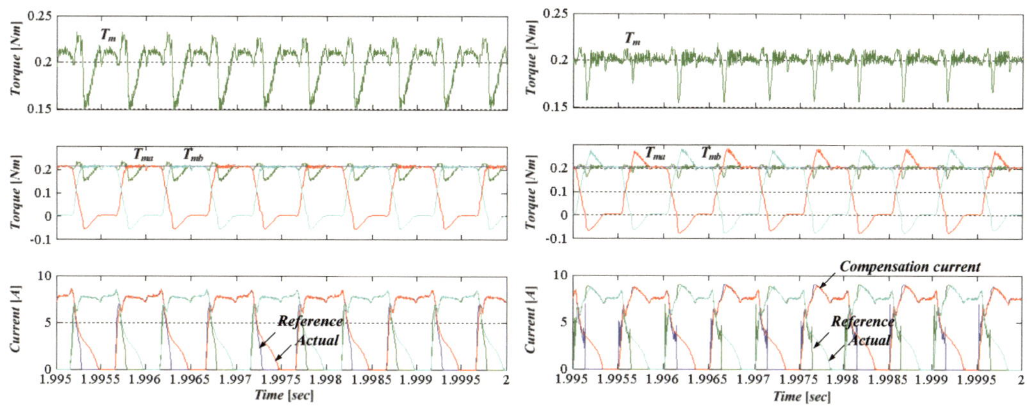

(a) Conventional cosine TSF (b) Proposed TSF control scheme

Fig. 30. Compared simulation result at 30,000[rpm].

Fig. 31 shows the compared experimental results according to the control schemes at 10,000[rpm] and no-load. With the conventional cosine TSF shown in Fig. 31(a), phase current can compensate the non-linear torque and constant torque can be achieved. In the proposed control scheme, the torque control performance is similar to the conventional cosine TSF. It is because that the outgoing phase current can be exhausted during the commutation region in the low speed and no load.

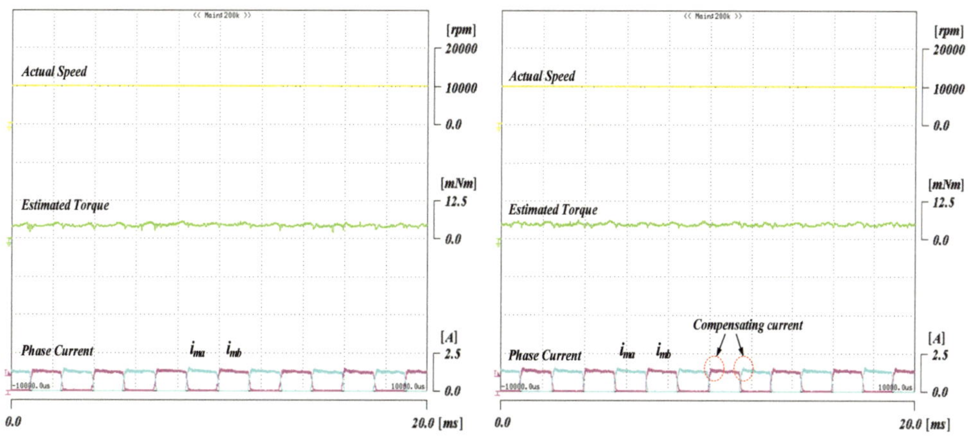

(a) Conventional cosine TSF (b) Proposed TSF control scheme

Fig. 31. Compared experimental results (At 10,000[rpm]).

Fig. 32 shows the experimental results at 30,000[rpm]. As shown in Fig. 32, the current of outgoing phase is not exhausted in the commutation region. And the tail current produces torque ripple as shown in Fig. 32(a). In the proposed control scheme, the incoming phase current can compensate the negative torque from tail current, and the torque ripple can be much decreased as shown in Fig. 32(b).

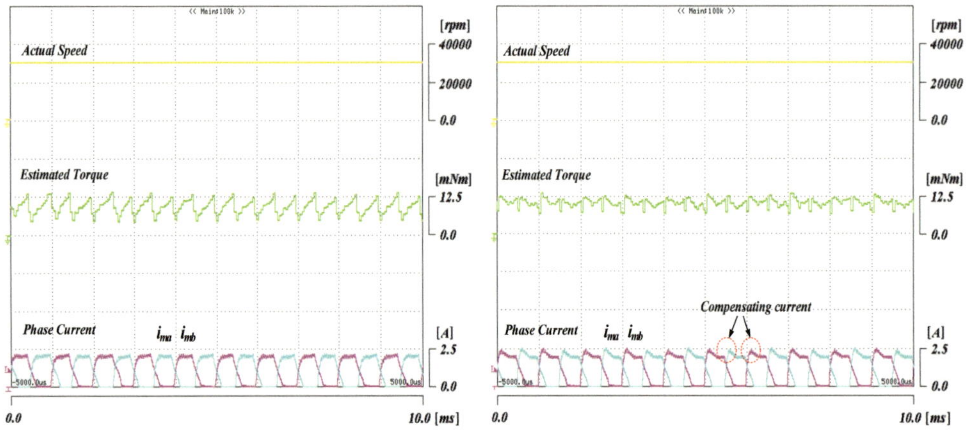

(a) Conventional cosine TSF (c) Proposed TSF control scheme

Fig. 32. Compared experimental results (At 30,000[rpm]).

Fig. 33 shows the experimental results with a practical heavy load fan. The fan load is increased according to the speed. In the 10,000[rpm], the tail current which is extended to the negative torque region produces high torque ripple as shown in Fig. 33(a). The compensation current can suppress the negative torque as shown in Fig. 33(b).

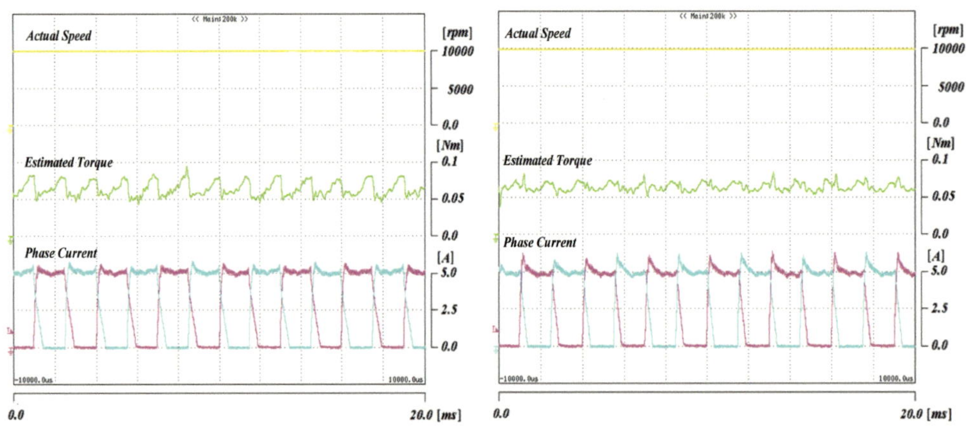

(a) Conventional cosine TSF (b) Proposed TSF control scheme

Fig. 33. Experimental results of heavy fan load (At 10,000[rpm]).

5. Conclusion

The reasons of the torque ripple in SRM are the inherent torque dead-zone from the design scheme, non-linear torque characteristics and the negative torque in a high speed region.
The inherent torque dead-zone can be removed by the design approach in two-phase SRM and Hybrid single-phase SRM. This torque ripple source cannot be removed by the control approach due to the inherent torque production mechanism of SRM.
Various torque control approaches can reduce the torque ripple from the non-linear torque characteristics of SRM. DTC and TSF method are good choice to improve the torque control

performance although they require a complex torque data of the practical motor. However, the performance of the current controller is much decreased. So, the conventional torque control scheme has torque ripple due to the current error and tail current. In order to reduce the tail current and current error in a high speed region, advance converter topologies which can supply additional boost voltage to increase the current control performance are introduced. These converters can reduce the current error in the torque control scheme, and it can improve the torque control performance in the high speed region. The negative torque from the tail current during the high speed can be compensated by the simple compensation technology. The detailed control scheme to compensate the negative torque is presented. The simulation and experimental results show the effectiveness the advanced converter topologies and torque compensation technology.

6. Acknowledgment

This work was supported by Energy Resource R&D program (2009T100100654) under the Ministry of Knowledge Economy, Republic of Korea.

7. References

A. K. Jain, N. Mohan, "SRM power converter for operation with high demagnetization voltage", Industry Applications, IEEE Transactions on, vol. 41, Issue 5, pp.1224-1231, Sept.-Oct. 2005.

A. Dahmane, F. Meebody, F.-M. Sargos, "A novel boost capacitor circuit to enhance the performance of the switched reluctance motor", Power Electronics Specialists Conference, 2001, pp.844-849, 17-21 Jun. 2001.

B.H. Bae, S. K. Sul, J.H. Kwon, J.S. Shin, "Implementation of sensorless vector control for super-high speed PMSM of turbo-compressor", in Proc. IEEE Industry Applications Conf., vol.2, Oct. 2001, pp.1203–1209.

C. H. Choi, S. H. Kim, Y. D. Kim, and K.H. Park, “A new torque control method of a switched reluctance motor using a torque-sharing function,” IEEE Trans. on Magnetics, vol. 38, no. 5, September 2002, pp. 3288-3290.

C. Sreekumar, V. Agarwal, "A Hybrid Control Algorithm for Voltage Regulation in DC–DC Boost Converter," IEEE Trans. on Industrial Electronics, vol. 55, no. 6, pp. 2530-2538, June 2008.

D.H. Lee, J. Liang, Z.G. Lee, and J. W. Ahn," A Simple Nonlinear Logical Torque Sharing Function for Low Torque Ripple SR Drive", Industrial Electronics, IEEE Transactions on, vol. 56, no.8, pp. 3021-3028, 2009.

D. H. Lee, J. W. Ahn, “A Novel Four-Level Converter and Instantaneous Switching Angle Detector for High Speed SRM Drive”, IEEE Trans. on Power Elctronics, Vol. 22, No. 5, pp. 2034-2041, 2007.

D. H. Lee, J. W. Ahn, “Design and Analysis of Hybrid Stator Bearingless SRM”, Journal of Electrical Engineering & Technology, Vol. 5, No. 4, pp. 571-579, Sep. 2010.

D. H. Lee, Lee Zhen-Guo, Liang Jianing, Ahn Jin-Woo, "Single-Phase SRM Drive With Torque Ripple Reduction and Power Factor Correction", IEEE Trans. on Industry Applications, Vol 43, Issue 6, Nov.-dec. 2007 pp.1578 - 1587

D. S. Schramm, B. W. Williams, and T. C. Green, "Torque ripple reduction of switched reluctance motors by phase current optimal profiling," in Proc. IEEE-PESC Conf. Rec.'92, 29 June-3 July 1992, pp. 857-860.

E. E. Kharashi, "Design and Analysis of Rolled Rotor Switched Reluctance Motor", Journal of Electrical Engineering and Technology, vol. 1, no. 4, pp. 472-481, 2006.

G. Dessouky, B. W. Williams, and J. E. Fletcher, "A novel power converter with voltage boosting capacitors for a four-phase SRM drive," IEEE Trans. Ind. Electron., vol. 45, no. 5, pp. 815-823, Oct. 1998.

H. Hannoun, M. Hilairet, C. Marchand, "Design of an SRM Speed Control Strategy for a Wide Range of Operating Speeds", IEEE Trans. on Industrial Electronics, Vol. 57, No. 9. Pp. 2911-2921, 2010.

I. Husain, "Minimization of torque ripple in SRM drives," IEEE Trans. on Industrial Electronics, vol. 49, no. 1, Feb. 2002, pp. 28-39.

I. Husain, M. Ehsani, "Torque ripple minimization in switched reluctance motor drives by PWM current control", IEEE Trans. on Power Electronics, vol. 11, no. 1, January 1996, pp. 83-88.

I.R. Kartono, K. Kajiwara, H. Dohmeki, "Dynamic simulation of maximizing the starting torque for super-high-speed drive of a 4/2 Switched Reluctance Motor", IEEE International Conference of Electrical Machines and System, pp.1-6, Sept. 2008.

J. C. Moreira, "Torque ripple minimization in switched reluctance motors via bi-cubic spline interpolation," in Proc. IEEE-PESC Conf. Rec.'92, 29 June-3 July 1992, pp. 851-856.

J. D. Lewis, H. R. Bolton, and N. W. Phillips, "Performance enhancement of single and two phase SR drives using a capacitor boost circuit," in European Power Electronics and Applications Conf. Rec., vol. 3, pp. 229–232, 1995.

J. F. Pan, N. C. Cheung, W. C. Gan, S. W. Zhao, "A Novel planar switched reluctance motor for industrial applications", IEEE Trans. on Magnetics, vol. 42, no. 10, Oct. 2006, pp.2836 - 2839.

J. Liang, D.H. Lee, J.W. Ahn, "A novel 4-level converter and instantaneous switching angle detector for high speed SR drive", in Proc. IEEE Power Electronics Specialists Conf. Jun. 2006, pp.1478~1483.

J. Liang, D. H. Lee, and J. W. Ahn, "Direct instantaneous torque control of switched reluctance machines using 4-level converters", Electric Power Applications, IET Transactions on, vol. 3, Issue 4, pp.313-323, Jul. 2009.

J. W. Ahn, S. J. Park, and D. H. Lee, "Hybrid excitation of SRM for reduction of vibration and acoustic noise," IEEE Trans. on Industrial Electronics, vol. 51, no. 2, April 2004, pp. 374-380.

J. W. Ahn, Sung-Jun Park, Dong-Hee Lee, "Novel encoder for switching angle control of SRM", IEEE Trans. on Industrial Electronics, vol. 53, no. 3, June 2006, pp. 848-853.

J. W. Ahn, T. H. Kim, D. H. Lee, "Performances of SRM for LSEV," Journal of Power Electronics, vol.5, no.1, Jan. 2005, pp. 45-54.

Kouta Kajiwara, Yong-Jae Kim, Hideo Dohmeki, "Analysis of the maximizing start torque of Switched Reluctance Motor for super high speed drive", Electrical Machines and Systems, ICEMS, pp.1428-1432 , Oct. 2007.

K. I. Hwu, C. M. Liaw, " DC-link voltage boosting and switching control for switched reluctance motor drives", Electric Power Applications, vol. 147, Issue 5, pp. 337-344, Sept. 2000.

K. Ohyama, M. Naguib, F. Nashed, K. Aso, H. Fujii, H. Uehara, "Design using finite element analysis of a switched reluctance motor for electric vehicle," Journal of Power Electronics, vol.6, no.2, April 2006, pp. 163-171.

K. Xin, Q. Zhan, J. Luo, "A new simple sensorless control method for switched reluctance motor drives", Journal of Electrical Engineering & Technology, vol. 1, no. 1, March, 2006. pp. 52-57.

L. Xu, C. Wang, "Accurate Rotor Position Detection and Sensorless Control of SRM for super-high Operation", IEEE Trans. on Power Electronics, Vol. 17, No. 5, pp. 757-763, 2002.

M. Dahmane, F.M. Tabar, F.M. Sargos, "An adapted converter for switched reluctance motor/generator for high speed applications", in Record, IEEE Industry Applications Conf. vol. 3, Oct. 2000, pp.1547-1554.

M. N. F. Nashed, K. Ohyama, K. Aso, H. Fujii, H. Uehara, "Automatic Turn-off Angle control for High Speed SRM Drives", Journal of Power Electronics, vol. 2, no. 1, pp. 81-88, 2007.

M. Krishnamurthy, C. S. Edrington, A. Emadi, P. Asadi, M. Ehsani, B. Fahimi, "Making the case for applications of switched reluctance motor technology in automotive products", IEEE Trans. on Power Electronics, vol. 21, no. 3, May 2006, pp. 659-675.

M.A. Rahman, A. Chiba, T. Fukao, "Super high speed electrical machines-summary", in Proc. Power Engineering Society General Meeting IEEE 6-10, vol.2, Jun. 2004, pp. 1272–1275.

N. Bianchi, S. Bolognani, F. Luise, "High speed drive using a slotless PM motor", IEEE Transactions on Power Electronics, vol.21, no.4, pp. 1083- 1090, July 2006.

R. C. Kavanagh, J. M. D. Murphy, and M. G. Egan, "Torque ripple minimization in switched reluctance drives using self-learning techniques," in Proc. IEEE-IECON Conf. Rec.'91, 28 Oct.-1 Nov. 1991, pp. 289-294.

R. Krishnan, Switched Reluctance Motor Drives: Modeling, Simulation, Analysis, Design, and Applications, CRC Press, 2001

R. Krishnan, D. Blanding, A. Bhanot, A. M. Staley, N. S. Lobo, "High reliability SRM drive system for aerospace applications.", in Proc. Industrial Electronics Society(IECON '03), vol. 2, Nov. 2003, pp. 1110-1115.

S. A. Bortoff, R. R. Kohan, R. Milman, "Adaptive Control of Variable Reluctance Motors : a Spline Function Approach", IEEE Trans. on Industrial Electronics, Vol. 45, No. 3, pp. 433-444, 1998.

S. Chan, H. R. Bolton, "Performance enhancement of single-phase switched-reluctance motor by DC link voltage boosting", Electric Power Applications, IEE Proceedings, Issue 5, Vol.2, pp.316-322, Sept. 1993

S. H. Won, J. H. Choi, J. Lee, "Windage Loss Reduction of High-Speed SRM Using Rotor Magnetic Saturation", IEEE Trans. on Magnetics, Vol. 44, No. 11, pp. 4147-4150, 2008.

S. I. Nabeta, I. E. Chabu, L. Lebensztajn, D. A. P. Correa, W. M. da Silva, ; K. Hameyer, "Mitigation of the Torque Ripple of a Switched Reluctance Motor Through a Multiobjective Optimization", IEEE Transactions on Magnetics, Vol. 44, Issue 6, June 2008, pp.1018 - 1021

S. Kozuka, N. Tanabe, J. Asama, A. Chiba, "Basic characteristics of 150,000r/min switched reluctance motor drive", Power and Energy Society General Meeting - Conversion and Delivery of Electrical Energy in the 21st Century, pp.1-4, July 2008.

S. K. Mondal, S. N. Bhadra, S. N. Saxena, "Application of current-source converter for use of SRM drive in transportation area" in Proc. Power Electronics and Drive Systems Conf., vol. 2, May 1997, pp.708-713.

S. M. Lukic, A. Emadi, "State-Switching Control Technique for Switched Reluctance Motor Drives : Theory and Implementation", IEEE Trans. on Industrial Electronics, Vol. 57, No. 9, pp. 2932-2938, 2010.

S. Mir, I. Husain, and M. E. Elbuluk, "Energy-efficient c-dump converters for switched reluctance motors," IEEE Trans. Power Electron., vol. 12, no. 5, pp. 912–921, Sep. 1997.

T. Raminosoa, B. Blunier, D. Fodorean, A. Miraoui, "Design and Optimization of a Switched Reluctance Motor Driving a Compressor for a PEM Fuel-Cell System for Automotive Applications", IEEE Trans. on Industrial Electronics, Vol. 57, No. 9, pp. 2988-2997, 2010.

T. Genda, H. Dohmeki, "Characteristics of 4/2 Switched Reluctance Motor for a high speed drive by the excitation angle", International Conference on Electrical Machines and Systems, ICEMS 15-18, pp.1-6, Nov 2009.

V. P. Vujicic, S. N. Vukosavic, M. B. Jovanovic, "Asymmetrical switched reluctance motor for a wide constant power range", IEEE Trans. on Energy Conversion, vol. 21, no. 1, Mar. 2006, pp.44-51.

W.L. Soong, G.B. Kliman, R. N. Johnson, R.A. White, . J.E. Miller, "Novel high speed induction motor for a commercial centrifugal compressor", in Proc. IEEE IAS Annual Meeting, vol.1, Oct. 1999, pp. 494-501.

X. D. Xue, K. W. E. Cheng, N. C. Cheung, "Multi-Objective Optimization Design of In-Wheel Switched Reluctance Motors in Electric Vehicles", IEEE Trans. on Industrial Electronics, Vol. 57, No. 9, pp. 2980-2987, 2010.

Y. H. Yoon, Y. C. Kim, S. H. Song, C. Y. Won, "Control of C-dump Converters fed from Switched Reluctance Motors on an Automotive Application," Journal of Power Electronics, vol.5, no.2, April 2005, pp. 120-128.

Y. K. Choi, H. S. Yoon, C. S. Koh, "Pole-Shape Optimization of a Switched-Reluctance Motor for Torque Ripple Reduction", IEEE Trans. on Magnetics, Vol. 43, Issue 4, April 2007, pp.1797 - 1800